Christoph Walther

Low Frequency and Circuit based Quantum Cascade Lasers

Christoph Walther

Low Frequency and Circuit based Quantum Cascade Lasers

Recent progress in Terahertz Quantum Cascade Lasers

Südwestdeutscher Verlag für Hochschulschriften

Impressum/Imprint (nur für Deutschland/only for Germany)
Bibliografische Information der Deutschen Nationalbibliothek: Die Deutsche Nationalbibliothek verzeichnet diese Publikation in der Deutschen Nationalbibliografie; detaillierte bibliografische Daten sind im Internet über http://dnb.d-nb.de abrufbar.
Alle in diesem Buch genannten Marken und Produktnamen unterliegen warenzeichen-, marken- oder patentrechtlichem Schutz bzw. sind Warenzeichen oder eingetragene Warenzeichen der jeweiligen Inhaber. Die Wiedergabe von Marken, Produktnamen, Gebrauchsnamen, Handelsnamen, Warenbezeichnungen u.s.w. in diesem Werk berechtigt auch ohne besondere Kennzeichnung nicht zu der Annahme, dass solche Namen im Sinne der Warenzeichen- und Markenschutzgesetzgebung als frei zu betrachten wären und daher von jedermann benutzt werden dürften.

Verlag: Südwestdeutscher Verlag für Hochschulschriften GmbH & Co. KG
Dudweiler Landstr. 99, 66123 Saarbrücken, Deutschland
Telefon +49 681 37 20 271-1, Telefax +49 681 37 20 271-0
Email: info@svh-verlag.de

Approved by: Zürich, ETH Zürich, Diss., 2011

Herstellung in Deutschland:
Schaltungsdienst Lange o.H.G., Berlin
Books on Demand GmbH, Norderstedt
Reha GmbH, Saarbrücken
Amazon Distribution GmbH, Leipzig
ISBN: 978-3-8381-2857-3

Imprint (only for USA, GB)
Bibliographic information published by the Deutsche Nationalbibliothek: The Deutsche Nationalbibliothek lists this publication in the Deutsche Nationalbibliografie; detailed bibliographic data are available in the Internet at http://dnb.d-nb.de.
Any brand names and product names mentioned in this book are subject to trademark, brand or patent protection and are trademarks or registered trademarks of their respective holders. The use of brand names, product names, common names, trade names, product descriptions etc. even without a particular marking in this works is in no way to be construed to mean that such names may be regarded as unrestricted in respect of trademark and brand protection legislation and could thus be used by anyone.

Publisher: Südwestdeutscher Verlag für Hochschulschriften GmbH & Co. KG
Dudweiler Landstr. 99, 66123 Saarbrücken, Germany
Phone +49 681 37 20 271-1, Fax +49 681 37 20 271-0
Email: info@svh-verlag.de

Printed in the U.S.A.
Printed in the U.K. by (see last page)
ISBN: 978-3-8381-2857-3

Dedicated to my wife, Marianik

Contents

Chapter 1

Introduction

1.1 The terahertz range

The creation and propagation of electromagnetic waves is described by the Maxwell equations. From a theoretical point of view, the frequency is 'just' a parameter that determines the photon energy of an electromagnetic wave. However the interaction of radiation with matter depends strongly on the frequency. The electromagnetic spectrum is devided in different regions corresponding to typical properties for the interaction with matter.

The frequency range from 0.3 to 10 THz is loosely defined as the terahertz range, corresponding to a wavelength of 30 - 1000 μm. In this frequency range dielectrics and dry materials such as paper, plastic, tissues and ceramics are fairly transparent to terahertz radiation, in contrast to conducting materials such as metals that are strong absorbers. Molecules in the gas phase, such as biomolecules, explosives, narcotics, and including water, have strong absorption lines in the terahertz range due to their rotational modes. The plasma frequency of doped semiconductors, the gap energy of high-temperature superconductors, optical phonons and transitions between hole states lie in the terahertz range. Van der Waals bonding energies, such as in soft molecular crystals, molecular clusters, biomolecules and organic semiconductors also correspond to frequencies in the terahertz. The development of imaging and spectroscopy techniques in the terahertz range give rise to new applications and strongly stimulate the development of terahertz sources and technology.

1.2 Terahertz applications

1.2.1 Astronomy

The terahertz range is very important for astronomy [1]. The cosmic background radiation contains spectral and spatial information on very distant newly formed galaxies, and on the early stages of star formation within gas clouds of our own galaxy. Examination of the spectral energy distributions in observable galaxies indicate that approximately one-half of the total luminosity and 98 % of the photons emitted since the big-bang fall into the terahertz range. Interstellar dust clouds may emit some 40'000 individual spectral lines, only a few thousand of which have been resolved and many of these have not been identified [2]. Spectral lines from the early universe appear strongly in the terahertz where they are less obscured by dust that often hides our view of galactic centers. Individual emission lines such as C^+ at 1.9 THz, the brightest line in the Milky Way terahertz spectrum, provide a detailed look at star forming regions. Many other abundant molecules, e.g., water, oxygen, carbon monoxide, nitrogen, and others can be probed in the terahertz regime. Since these signals are obscured from most earth-based observations, except from a very few high-altitude observatories, aircraft, or balloon platforms, they provide strong motivation for a number of existing or upcoming space astrophysics instruments. For the high resolution detection of such emission lines, highly sensitive heterodyne receivers are developed. The signal coming from the sky is mixed with a local oscillator to get a signal in the gigahertz range. Requirements for terahertz local oscillators are cw operation, frequency stability, narrowband, tunability, a compact size and a low power consumption.

1.2.2 Spectroscopy

The advantage of terahertz spectroscopy is in the strength of the absorption lines of rotational and vibrational excitations of many molecules that peak in this frequency range [1]. Modern applications foreseen for terahertz spectroscopy require rapid scan and gas identification systems, and optical pulse terahertz time-domain spectroscopy. Many interesting molecules have unique spectral fingerprints in the terahertz range that allow their identification. Intermolecular vibrations in some chemicals and organic molecules appear in the terahertz and allow to study the dynamics of large biomolecules [3]. DNA identification and bio-sensing applications are investigated as well. Security applications are considered since illegal narcotic substances and several types of explosives have fingerprints in the terahertz region [4], whereas packaging and clothing are often transparent for terahertz radiation.

1.2.3 Imaging

Terahertz waves are transmitted through many materials that are opaque at optical frequencies and give much higher imaging resolution than microwaves. X-rays provide often very low contrast for such materials. Terahertz systems are ideal for imaging dry dielectric substances including paper, plastics, ceramics, and many others. These materials are relatively transparent in the terahertz range. Terahertz imaging is also an attractive tool for the medical community for noninvasive sampling and for nondestructive probing of biological materials, such as tissues. Other potential applications are security screening and manufacturing quality control. In coherent terahertz imaging [5], measurement of the transmitted or reflected terahertz energy incident on a sample is processed to reveal spectral content, signal strength and time of flight data. The latter can be used to determine the refractive index, amplitude and phase, and sample thickness. The principle of coherent terahertz imaging involves the generation and detection of terahertz electromagnetic transients that are produced in a photoconductor or a crystal by intense femtosecond optical laser pulses. Scanning either the terahertz generator or the sample itself allows a 2D image to be built up over time.

1.3 Terahertz sources

In terms of sources, the terahertz range is an underdeveloped part of the electromagnetic spectrum, which is related to the difficulty of generating terahertz radiation. Only thermal sources represent a simple but inefficient way to produce terahertz radiation. Since the development of the radar, especially during world war II and the cold war, there has been a strong military interest for efficient radiation sources at high microwave frequencies or even terahertz frequencies. For years the scientific activity in the terahertz range was limited to molecular spectroscopy using thermal sources. Other frequency regions were in the focus of scientists, as a short historical review shows.

In 1865, Maxwell modified Ampère's law to complete the set of four equations which bear since then his name. It was Maxwell's prediction that light was an electromagnetic wave phenomenon, and that electromagnetic waves of all frequencies could be produced. Those predictions drew the attention of all physicists and stimulated much theoretical and experimental research into electromagnetism during the last part of the nineteenth century. Shortly after the demonstration of Maxwell's equations, Hertz generated, propagated and detected radio waves in his laboratory as early as 1887, based on a spark gap generator. In 1912 Armstrong demonstrated the first electronic oscillator based on a triode vacuum tube, a very important achievement in the field of electronics. Einstein established the foundations of the laser and maser in 1916 by the formulation of the stimulated and spontaneous emission and their relationship to the radiation absorption process in atomic and molecular systems. It was only in 1954 that Townes invented the microwave maser, the first quantum electronic oscillator. In 1958 Schawlow and Towns, working at Bell Labs, filed a patent application for a proposed "optical maser" and published their theoretical calculations. Finally in 1960 the first optical oscillator was demonstrated with the ruby laser by Maiman. The field of modern optics was born. Since then, electronics and optics developed as two independent disciplines for the generation and manipulation of electromagnetic waves. While electronic sources were mainly developed in the low frequency range (< 100 GHz) of the electromagnetic spectrum, optical sources were initially developed at high frequencies, such as $200 - 500$ THz. Nowadays, wide parts of the electromagnetic spectrum ranging from radio waves

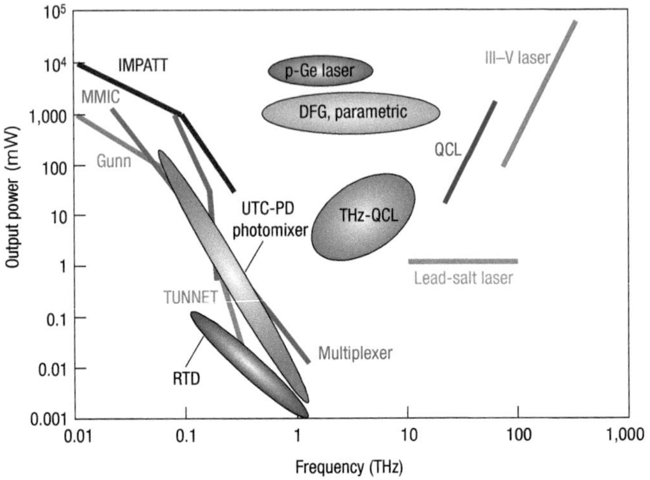

Figure 1.1: (reprinted from Ref. [3]) Emission power of various terahertz sources. Solid lines are for conventional terahertz sources. Ovals denote recent terahertz sources.

to X-rays are covered with sources of coherent or narrowband radiation due to the technological progress and intense research in electronics and optics. The terahertz range remains as one of the last 'frontiers'. While the efficiency of terahertz generation by high frequency electronics decreases with frequency, traditional optical sources such as III-V lasers or lead salt lasers have a low-frequency cutoff, shown in Fig. 1.1. New sources are promising in terms of powers but so far non of them is simultaneously compact, efficient, broadly tunable, and works at room temperature. A room temperature terahertz source comparable to an electronic oscillator or a laser diode is still inexistent and a remaining challenge. The main terahertz sources are summarized in the following paragraphs. Excellent reviews on terahertz technology are found in [1, 3, 6].

1.3.1 Electronic sources

Microwave upconverters

Microwave oscillators are based on diodes with a region of negative differential resistance, that acts as a broadband amplifier medium, in combination with an optical resonator, required for feed-back. Gunn diodes, impact ionization avalanche transit-time (IMPATT), tunneling transit-time (TUNNETT) and resonant tunneling diodes (RTD) are some examples. RTDs have the highest oscillations frequencies among these diodes [7] and fundamental oscillations up to 712 GHz [8] have been demonstrated in 1991(!). The power gain cutoff frequency f_{max} of InP heterojunction bipolar transistors is increased steadily through scaling of the lithographic feature size in fabrication. The highest value reported for f_{max} is 755 GHz at 250 nm feature size [9]. Monolithic microwave integrated circuit (MMIC) power-amplifier chips have appeared and promise to extend baseband frequency coverage up to at least 200 GHz. Oscillations in in the terahertz range are generated by up-convertion of a fundamental microwave oscillator. The non-linearity of diodes is exploited to generate harmonics. The diode is mounted in a single-mode waveguide that filters the selected harmonics. The conversion efficiency of frequency multipliers decreases rapidly for higher order harmonics. Therefore a succession of frequency doublers and triplers are used, with a typical conversion efficiency of 5-10% per stage. A high power fundamental oscillator is required to compensate the conversion losses. Typically a few μW of coherent single mode power at 2 THz are generated after such a multiplier chain when starting with a base oscillator of 100 mW at 100 GHz. Multiplier chains are commercially available [10].

Backward wave oscillator

Backward wave oscillators (BWOs) are vacuum tubes developed for the generation of intense microwave power. Accelerated electrons are modulated in velocity by a periodic field along their trajectory to the anode and hence transform their energy to electromagnetic radiation that is highly monochromatic and coherent. An external magnetic field is required for the collimation of the electron beam. BWOs work in the spectral range of 30 GHz to 1.5 THz [11]. The output power strongly depends on the frequency. Typical powers are 10 mW, but it may exceed 100 mW [12]. By adjusting the acceleration voltage, the BWO can be tuned over a frequency range of about a factor of two. BWOs for frequencies below 180 GHz are compact and air-cooled. At higher frequency, the required magnetic field increases, and requires large magnets with a water-cooling system. The typical weight of a BWO working in the terahertz is more than 15 kg.

1.3.2 Optical sources

Photomixing

Two cw-lasers with slightly different frequencies are focussed onto a small area of an appropriate photoconductor with sub-picosecond lifetimes, such as low temperature grown Gallium Arsenide, to generate carriers between closely spaced and biased electrodes printed on the semiconductor. The laser induced photocarriers short the gap, producing a photocurrent, which is modulated at the laser difference frequency. This current is coupled to an RF circuit or antenna that couples out or radiates the terahertz energy. Typical optical to terahertz conversion efficiencies are below 10^{-5}.

Among various types of long-wavelength photodiode technologies, a uni-traveling-carrier photodiode (UTC-PD) has exhibited the highest output power at frequencies from 100 GHz to 1 THz. Maximum cw output powers of 20 mW, respectively 10 μW, have been achieved at 100 GHz, respectively 1 THz, using laser diodes operating at 1.55 μm [13].

Femtosecond pulse terahertz generation

A similar setup is used for the generation of broadband terahertz energy. Instead of two cw lasers, a short pulse femtosecond optical laser, such as a Ti:Sapphire laser is used to illuminate the gap between closely spaced electrodes on a photoconductor generating carriers which are then accelerated in an applied field [1]. The resulting current pulse, which is coupled to an RF antenna, has frequency components that reflect the pulse duration. Typically the spectra has a frequency content from $0.2-2$ THz. The average power level over the entire spectrum is nanowatts to microwatts.

Nonlinear optics

THz waves are generated by nonlinear optical effects such as optical rectification (limited to femtosecond laser excitation), difference-frequency generation (DFG) or optical parametric oscillation [3]. A terahertz parametric generator (TPG) consists of a high power laser (Q-switched Nd:YAG) that is used to illuminate a crystal causing optical parametric oscillations via the crystal nonlinearity and the polariton mode scattering. The parametric process creates a near-IR photon (idler) close in wavelength to that of the pump and a terahertz difference photon. The spectrum contains a wide range of terahertz frequencies. Narrowband terahertz emission is obtained using an optical resonator for the idler to form a terahertz parametric oscillator (TPO), or by stimulating a specific frequency of the idler wave in a process called injection seeding (IS). An IS-TPG, that can produce a linewidth of less than 100 MHz and frequency tunable from 0.7 to 3 THz with terahertz peak powers exceeding 100 mW has been demonstrated [14].

Terahertz lasers

Gas lasers are used for terahertz generation at frequencies from 0.9 to 3 THz with cw output powers in the range of $1 - 30$ mW. A gas laser consists of a low pressure flowing gas cavity containing a particular gas such as CH_4, N_2 and others. A carbon dioxide laser is used for optical pumping of the gas whose emission line dictates the lasing frequency. Gas lasers have no tunability and are very large, with dimensions exceeding 2.5 m in length. However, a miniaturized version of a gas laser was recently reported, which delivers 30 mW at 2.5 THz, its dimensions are 75 x 30 x 10 cm and the weight is 20 kg [15].

In p-Germanium semiconductor lasers are used for generation of terahertz radiation. A population inversion between light hole (LH) and heavy hole (HH) bands of a p-doped Germanium crystal is induced by perpendicular electric and magnetic fields. The electric and magnetic fields accelerate the HHs above the optical phonon energy; part of them are scattered in the LH band where LHs are accumulated on closed paths just below optical phonon energy. A key feature of the pe-Ge laser is the large continuous tuning range of up to $1 - 4$ THz by tuning the magnetic field. The p-Ge laser has a low efficiency, measured efficiencies range from 10^{-6} to

10^{-4} [16], works only in pulsed mode operation with a maximum duty cycle of a few percents, and is cryogenically cooled to 4 K. Recently, peak output powers of 40 W in the 1.5 – 4.2 THz range have been demonstrated [17].

Terahertz quantum cascade lasers (QCLs) are compact solid state sources of coherent radiation. Power levels of tens of milliwatts are commonly achieved, the highest reported peak power being 248 mW in pulsed mode and 138 mW in continuous mode operation at 4.4 THz [18]. Continuous tuning over 3.6% of the center frequency is achieved on a terahertz wire laser at 3.8 THz [19]. So far, terahertz QCLs are limited to cryogenic operation temperatures, the highest operation temperature being 186 K [20].

1.4 Bridging the gap

In conventional gas lasers, respectively semiconductor interband lasers, the lasing frequency is fixed by the atomic or molecular transition, respectively the bandgap. QCLs are unipolar semiconductor heterostructure lasers based on intersubband transitions in quantum wells. The frequency of the optical transition is not fixed by the material properties but is designed through the layer sequence of the heterostructure. A priori any frequency can be engineered. The frequency range of QCLs spans impressive two orders of magnitudes, ranging from the near infrared (above 100 THz) to the terahertz region. The first part of this thesis focusses on the development of low frequency terahertz QCLs below 2 THz. A design scheme is developed that faces the challenges of the low transition frequency. The main result are a series of QCLs that cover the frequency range below 2 THz and set

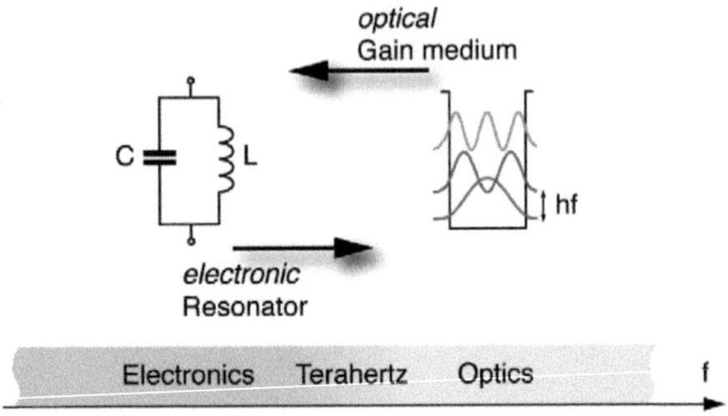

Figure 1.2: The scope of the thesis is to bridge the terahertz gap. In the first part an optical gain medium is pushed as far as possible towards electronic frequencies. In the second part an electronic resonator is brought to the optical regime and combined with an optical active region to form a hybrid laser-oscillator.

a record for the lowest operation frequency of 1.2 THz (without strong magnetic field). Through the comparison of different lasers and magneto-transport measurements, the limiting mechanism for lasing operation at low frequency is discussed. From a conceptional point of view, an optical gain medium, the QCL active region, is pushed as far as possible to the electronic regime of the electromagnetic spectrum.

In the second part of this thesis, the opposite is pursued, as shown in Fig. 1.2. An electronic concept, e.g. a circuit based resonator comprising an inductor and a capacitor is taken to the optical regime. It is combined with an optical active gain medium to form a hybrid laser-oscillator, that consists of an optical amplifier through stimulated emission of radiation, but an electronic resonator for feedback. The optical field is not a standing wave that results from bouncing back and forth in an optical cavity, but its energy is oscillating between the electric field energy in the capacitor and magnetic field energy in the inductor. The circuit based laser has the property of being a deep sub-wavelength sized microcavity laser. The effective mode volume is among the smallest for electrically pumped lasers. The large electric field generated by vacuum fluctuations in the circuit based resonator enables the observation of cavity quantum electrodynamics effects such as the strong-coupling regime of the light-matter interaction or the enhancement of spontaneous emission, the so called Purcell effect. From the emission characteristics of the circuit based laser a strong Purcell effect is deduced. Intersubband polaritons are demonstrated using an appropriate active region in the circuit based resonator. The circuit based resonator in combination with an active region could lead to a class of new devices to generate and manipulate terahertz radiation that exploit cavity quantum electrodynamic effects.

Chapter 2

Theory

2.1 The QCL, a unipolar intersubband laser

The QCL is based on intersubband transitions in contrast to other semiconductor lasers that are based on interband transitions. Fig. 2.1 shows a comparison of an interband and intersubband transition in a quantum well. In lasers based on interband transitions, conduction band electrons and valance band holes radiatively recombine across the bandgap. The bandgap essentially determines the emission wavelength. The width of the gain spectrum has a strong temperature dependence through the hole and electron distribution according to Fermi's statistic. In contrast, the QCL is a unipolar intersubband laser. It relies on only one type of carriers making electronic transitions between conduction band states (subbands) arising from size quantization in a semiconductor heterostructure. These transitions are denoted as intersubband transitions. The initial and final subband has the same curvature and therefore the joint density of states is atomic like, leading to narrow gain [22] without a direct temperature dependance of the linewidth. The

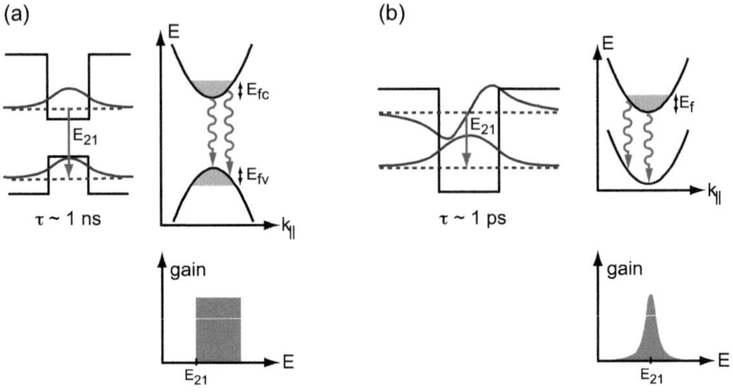

Figure 2.1: (Figure reprinted from [21]). (a): Interband transition between the conduction and valence band in a quantum well leads to a flat gain curve. (b): Intersubband transitions in the conduction band have an atomic like joint density of states; the gain curve has a Lorentzian shape.

transition energy is no longer fixed by the bandgap of the material, but can be engineered by the quantum design of the heterostructure. The lifetimes of intersubband transitions are typically three orders of magnitude shorter than those of interband transitions.

The idea of a unipolar laser based on transitions between states belonging to the same band can be traced to the original proposal of B. Lax in 1960 of a laser based on an inversion between magnetic Landau levels in a solid [23]. The advent of molecular beam epitaxy in the late 1960's enabled for the first time the fabrication of heterostructures based on III-V compounds with very sharp interfaces and excellent compositional control [24]. The seminal work of Kazarinov and Suris in 1971 represents the first proposal to use intersubband transitions in quantum wells for light amplification [25]. Esaki and Tsu investigated the transport in multi-quantum well structures and observed resonant tunneling on a double barrier diode in 1974

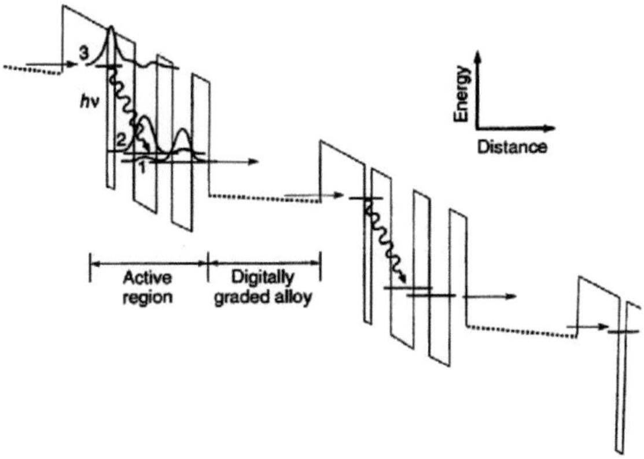

Figure 2.2: Energy band diagram of the first QCL reprinted from Ref. [29]. The dashed lines are the effective conduction band edges of the digitally graded electron relaxation/injecting regions. The moduli squared of the relevant wave functions are shown. The wavy arrow indicates the laser transition. The photon energy is $h\nu = E_3 - E_2 = 295$ meV.

[26]. The first observation of intersubband absorption in a GaAs/AlGaAs multi-quantum well dates from 1985 by West and Eglash [27], followed by the observation of intersubband luminescence by resonant tunneling in 1989 by Helm [28]. A large number of proposals for intersubband lasers came up in the 1980's and early 1990's. The first QCL was demonstrated in 1994 by Faist [29].

A QCL consists of a cascaded repetition (typically tens to hundreds) of an active region sequence. Fig. 2.2 shows the bandstructure of the first QCL. Electrons are injected through a thick barrier into the $n = 3$ energy subband of the active region by resonant tunneling. The reduced spatial overlap between the $n = 3$ and $n = 2$ subbands and the strong coupling of the latter to the $n = 1$ subband of the adjacent well ensure a population inversion between subbands 3 and 2. The energy difference $E_2 - E_1$ is close to the optical phonon energy and results in a short extraction time from the $n = 2$ subband. In the relaxation region, the electrons thermalize through scattering with other electrons and emission of optical phonons. From the injector electrons get re-injected into the upper subband of the following period. Assuming a unity injection efficiency (all the electrons are injected from the injector into the upper subband), the condition for a population inversion is $\tau_{32} > \tau_2$. Conduction electrons are provided through doping of the active region. Since scattering with ionized dopants broadens the optical transition [30], the injection/relaxation region is modulation doped.

After the demonstration of the first QCL, the active region design, waveguide losses and thermal management have shown an impressive improvement and led to the technologically important continuous wave operation of a mid-infrared QCL at room temperature in 2002 [31]. In the same year the first terahertz QCL, operating at 4.4 THz, has been demonstrated [32]. Nowadays, QCLs are sold commercially [33, 34, 35] and reach high performance at room temperature in the 3 to 20 μm range. At 4.6 μm Watt levels in continuous wave operation have been demonstrated at room temperature with wall plug efficiencies higher than 10% [36, 37]. The wavelength coverage of QCLs spans two decades from the near telecom range at 2.63 μm (114 THz) to 250 μm (1.2 THz) in the terahertz region [38, 39]. The restrahlen band at ~34 μm (~9 THz) (in GaAs and InP) is characterized by very high losses due to longitudinal optical phonons. QCL lasers operating above the reststrahlen band are called mid-infrared QCLs, and those operating below are called far-infrared or terahertz QCLs.

2.2 QCL modeling

2.2.1 Energy states in heterostructures

The active region of a quantum cascade laser consists of a complex multi-quantum well heterostructure. The quantum wells are formed by the conduction band offset of an alternation of thin layers of two semiconductor materials. For the computation of the bandstructure, the envelope function formalism in the Kane approximation is used as described in [40]. The latter gives a good description of the bulk crystal close to the high-symmetry points in the Brillouin zone. The electronic wavefunctions of a III-V heterostructure are written as the product of a slowly varying envelope function by a rapidly varying (atomic-like) function. The latter can be assumed to be identical in the different III-V materials. Therefore the analysis of the heterostructure eigenstates $\psi(\mathbf{r})$ is carried out in terms of the products of atomic-like functions that are identical in both host materials and slowly varying functions at the scale of the host elementary cells that will alone experience the difference between the bandstructures of the host material.

$$\psi(\mathbf{r}) = \sum_{l} f_l^{(A,B)}(\mathbf{r})\, u_{l,\mathbf{k_0}}(\mathbf{r}) \tag{2.2.1}$$

where the summation l is over the band edges. $f_l^{(A,B)}(\mathbf{r})$ is the envelope function, and $u_{l,\mathbf{k_0}}(\mathbf{r})$ the periodic part of the Bloch functions.

The whole bandstructure of the bulk material should be considered in order to obtain exact results within the envelope function approximation. However it has been shown that the 4-band Kane model which takes into account the full valence- and conduction-bandstructure works well in the cases of interest [41]. It is assumed that the in-plane momentum vanishes, so that the heavy-hole state is decoupled, and the remaining 3-band Hamiltonian takes into account the conduction, light-hole and split-off position dependent band edges. The 3-band model can be further simplified through a unitary transformation which allows to replace light-hole and split-off bands with an "effective" valence band v.

For the computation of the energy levels above the edge of the conduction band, the problem can be transformed to the solution of a Schroedinger-like equation with an energy and position-dependent effective mass

$$p_z \frac{1}{2m(E,z)} p_z \phi_c + E_c(z)\phi_c = E\phi_c \tag{2.2.2}$$

where the energy-dependent effective mass is simply given by

$$m(E,z) = m_0 \frac{E - E_v(z)}{E_p} \tag{2.2.3}$$

The parameters $E_c - E_v$ and E_p are determined from measured values of $m^* = m(E_c)$ of the constituent materials and from the knowledge of their nonparabolicity coefficient, defined as $\gamma^{-1} = 2m^*|E_c - E_v|/\hbar^2$ [42]. The experimental values of m^* and γ are in principle determined by the whole band structure, and they could therefore introduce phenomenologically the effect of remote bands in the Kane formulation.

The solutions of equation 2.2.2 don't form an orthogonal set of functions, so that dipole matrix elements between them are ill-defined. The solution to this problem is to go back to the two-band model and compute the matrix element including the valence band part. The dipole matrix element now reads

$$z_{ij} = \frac{\hbar}{2(E_j - E_i)} \left\langle \phi_i \left| p_z \frac{m_0}{m(E_i, z)} + \frac{m_0}{m(E_j, z)} p_z \right| \phi_j \right\rangle \tag{2.2.4}$$

where the momentum operator is defined as $p_z = -i\hbar(\partial/\partial_z)$. This the heart of the $\mathbf{k} \cdot \mathbf{p}$ approximation: the same matrix element (p) defines both the gap and the interaction with the optical wave.

2.2.2 Spontaneous and stimulated emission

The interaction of a quantum system with radiation is extensively described in Ref. [43, 44]. Spontaneous and stimulated emission rates can be derived for an ensemble of atomic two level systems that weakly interacts with a radiation field using Fermi's golden rule. The spontaneous emission rate W_{sp}^{21} in free space or in a homogenous material with refractive index n_{op} is obtained considering the coupling of the 2 level systems to all vacuum modes

$$W_{sp}^{21} = \frac{q^2 r_{12}^2 \omega_0^3 n_{op}}{3\pi c^3 \hbar \varepsilon_0} = \frac{1}{\tau_{sp}} \tag{2.2.5}$$

where $\hbar\omega_0 = E_2 - E_1$ is the photon energy, and r_{12} the dipole matrix element between the excited state 2 and the ground state 1 of the two level system. The stimulated emission rate W_{st}^{21} due to the interaction of a monochromatic traveling wave with an ensemble of two level systems is

$$W_{st}^{21}(\nu) = \frac{2\pi^2 q^2 I_\nu}{ch^2 n_{op}\varepsilon_0} \left\langle |\hat{\mathbf{e}} \cdot \mathbf{r}_{12}|^2 \right\rangle_{atoms} \mathcal{L}(\nu - \nu_0) \tag{2.2.6}$$

where $\left\langle |\hat{\mathbf{e}} \cdot \mathbf{r}_{12}|^2 \right\rangle_{atoms}$ takes into account the orientation of the dipole matrix element $\mathbf{r}_{12}$ of the atoms in respect to the polarization of the monochromatic wave, given by the unit vector $\hat{\mathbf{e}}$. I_ν is the electromagnetic wave intensity. To account for the broadening of the $2 \rightarrow 1$ transition the normalized lineshape function $\mathcal{L}(\nu - \nu_{21}) = (\Delta\nu/2\pi)/[(\nu - \nu_{21})^2 + (\Delta\nu/2)^2]$ is used, where $\Delta\nu$ is the full width at half maximum.

2.2.3 Gain in QCLs

The gain of an optical mode in a QCL active region is deduced using the result of the previous section. Gain is provided by an electronic population inversion between two subbands. For now the upper subband is 2 and the lower subband 1, and z_{12} is the dipole matrix element of the $2 \rightarrow 1$ transition. The coupling of the mode to the intersubband transition is described by the confinement factor Γ that is defined as the fraction of the electric field energy that couples to the intersubband transition.

$$\Gamma = \frac{\int_{V gain} \varepsilon(r) \left|E_z(r)\right|^2 d^3r}{\int_{V space} \varepsilon(r) \left|\mathbf{E}(r)\right|^2 d^3r} \tag{2.2.7}$$

where E_z is the component of the electric field that couples to the intersubband transition (TM polarization). First, the maximal gain in the material is deduced, corresponding to $\Gamma = 1$. The stimulated emitted power per unit volume is related to the gain coefficient by considering the increase of the intensity of the traveling wave that crosses the gain medium

$$W_{st}^{21}(\nu)\,\Delta N\,h\nu = \frac{dI_\nu(z)}{dz} = g(\nu)I_\nu(z) \tag{2.2.8}$$

where $\Delta N = N_2 - N_1$ and N_2 respectively N_1, is the volume electronic density in the upper, respectively lower subband. The gain is identified as

$$g(\nu) = \frac{2\pi^2 q^2 z_{12}^2}{n_{op}\varepsilon_0 \lambda h}\mathcal{L}(\nu - \nu_0)\Delta N \tag{2.2.9}$$

where $\left\langle |\hat{\mathbf{e}}\cdot\mathbf{r}_{12}|^2 \right\rangle_{atoms} = z_{12}^2$. Unlike in atomic systems, the dipoles in a QCL have all the same orientation. The peak material gain $g_p = g(\nu_0)$ is the maximal available gain in the material. The peak material gain writes:

$$g_p = \frac{4\pi q^2 z_{12}^2}{\varepsilon_0 n_{op}\lambda 2\gamma_{12} L_p}(n_2 - n_1) \tag{2.2.10}$$

Where $2\gamma_{12}$ is the full width at half maximum of the $2 \rightarrow 1$ transition in energy units, L_p the length of one period and $n_2 = N_2 L_p$, $n_1 = N_1 L_p$ are the sheet densities of the upper and lower subband. The maximal gain that a propagating mode experiences (e.g. in a laser cavity) is called modal gain defined by $g_m = \Gamma g_p = g_c \Delta n$, where g_c is the gain cross section and $\Delta n = n_2 - n_1$ is the sheet inversion density. The gain cross section is given by

$$g_c = \frac{\Gamma\, 4\pi q^2 z_{12}^2}{\varepsilon_0 n_{op}\lambda 2\gamma_{12} L_p} \tag{2.2.11}$$

The oscillator strength is commonly used to characterize the strength of an optical transition. It is defined by

$$f_{21} = \frac{2m_0}{\hbar^2} E_{21} z_{12}^2 \tag{2.2.12}$$

m_0 is the electron mass. If non-parabolicity can be neglected, the oscillator strength satisfies a sum rule that is stated as

$$\sum_{j \neq i} f_{ji} = \frac{m_0}{m^*} \tag{2.2.13}$$

where m^* is the effective mass (for GaAs $m^* = 0.067m_0$). The modification to the sum rule for a spatial dependent effective mass and non-parabolicity is discussed in Ref. [41]. For low transition energies, as it will be the case of low frequency terahertz QCLs, the non-parabolicity can be neglected. The oscillator strength is proportional to the gain and allows for a simple comparison of the optical intersubband transition of different QCL designs. A consequence of the sum rule is that the oscillator strength from higher states may be significantly larger than from the ground state, since the oscillator strength of downward transitions is negative.

2.2.4 Rate equations

A rate equation model that takes one optical mode into account is used to derive the principle characteristics of the transport through the active region [22]. For simplicity the 3 state active region shown in Fig. 2.3 is considered for modeling.

The upper and lower state of the laser transition and the ground state of the injector n_g are used for the modeling. Rate equations are formulated for the sheet densities n_3 and n_2. The units of the sheet density are $1/\mathrm{m}^2$. The photon flux per period per stripe width is defined as $S = cpL_p/(n_{op}V)$ and has $1/\mathrm{ms}$ units, where p is the photon number in the lasing mode. The number of periods N_p is contained implicitly in this expression since $V/L_p = AN_p$, where A is the laser surface. The rate equations for the upper laser state 3, lower laser state 2 and the photon density S are given by

$$\frac{dn_3}{dt} = \frac{J}{q}\eta - \frac{n_3}{\tau_3} - Sg_c(n_3 - n_2) \tag{2.2.14}$$

$$\frac{dn_2}{dt} = \frac{J}{q}(1-\eta) + \frac{n_3}{\tau_{32}} + Sg_c(n_3 - n_2) - \frac{n_2 - n_2^{therm}}{\tau_2} \tag{2.2.15}$$

$$\frac{dS}{dt} = \left(\frac{c}{n}\right)\left\{[g_c(n_3 - n_2) - \alpha]S + \beta\frac{n_3}{\tau_{sp}}\right\} \tag{2.2.16}$$

where n_2^{therm} is the thermally backfilled electrons from the injector state. The sheet density of the injector is $n_g \gg n_3, n_2$. η is the injection efficiency into the upper state, g_c is the gain cross section, c the velocity of light in vacuum, n the mode refractive index, α are the total losses being the sum of mirror losses and waveguide losses. β is the spontaneous emission coupling factor, the ratio of spontaneously emitted photons in the lasing mode. While in a laser waveguide β is in the order of 0.001, it can get quite close to 1 in microcavities.

Assuming for the moment unity injection efficiency ($\eta = 1$). The threshold current density is obtained, by neglecting the spontaneous emission and solving the steady state rate equations with the condition $S \rightarrow 0$. The threshold current density is

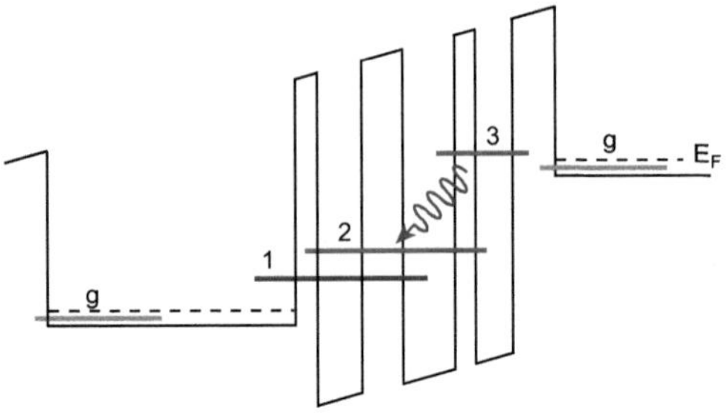

Figure 2.3: Schematic diagram of one period of a QCL active region. (3) is the upper state and (2) the lower state of the laser transition, and (g) is the lowest injector state. E_F is the quasi-Fermi level in the injector.

expressed as

$$J_{th} = q\frac{\alpha/g_c + n_2^{therm}}{\tau_3(1-\tau_2/\tau_{32})} \tag{2.2.17}$$

The condition for a population inversion is $\tau_{32} > \tau_2$. At cryogenic temperature < 100 K, backfilling can be ignored. In a similar way, the slope efficiency per facet dP/dI is deduced

$$\frac{dP}{dI} = N_p\frac{h\nu}{q}\frac{\alpha_m}{\alpha_{tot}}\frac{\tau_{eff}}{\tau_{eff}+\tau_2} \tag{2.2.18}$$

where N_p is the number of periods, α_m the mirror losses, α_{tot} the total losses and the effective lifetime τ_{eff} is defined by $\tau_{eff} = \tau_3\,(1-\tau_2/\tau_{32})$. The internal quantum efficiency for the photon generation above the threshold is defined as

$$\eta_{int} = \frac{\tau_{eff}}{\tau_{eff}+\tau_2} \tag{2.2.19}$$

Due to the finite lower state lifetime τ_2, the internal quantum efficiency is below unity.

At the laser threshold and above, the population inversion is clamped at its threshold value, resulting in a reduction of the upper state lifetime due to stimulated emission. The reduction leads to a discontinuity of the differential resistance at threshold [45]. Under the assumption that the upper state population n_3 is in a linear relationship with the applied voltage, the differential resistance dV/dJ is proportional to dn_3/dJ. Solving the rate equations above and below threshold, following relation is derived for the ratio of the differential resistance above and below threshold

$$R_{d,S>0}/R_{d,S=0} = \frac{\tau_2}{\tau_{eff} + \tau_2} \quad (2.2.20)$$

Eq. 2.2.20 is the topic of a controversy. The authors of Ref. [46] developed a three-level density-matrix model to describe the role of coherence in resonant tunneling transport of QCLs. They derived a different expression for the slope discontinuity at the lasing threshold. They claim that the discontinuity is a direct measure of the population inversion in contradiction of being indicative for the ratio of the laser level lifetimes. They argue that the subband populations cannot be assumed to be directly proportional to the bias. In Ref. [47] a similar four-level density-matrix model is developed, and in Fig. 5 of that reference are shown the computed populations. The upper state population is clearly in a linear relationship with the electric field when the injector couples to the upper state, confirming the assumption that was used to derive Eq. 2.2.20.

For a non-unitary injection efficiency following formulas are obtained for the threshold current density, slope efficiency and the ratio of the differential resistance:

$$J_{th} = \frac{q\alpha/g_c}{\tau_3(1-\tau_2/\tau_{32})\eta - \tau_2(1-\eta)} \tag{2.2.21}$$

$$\frac{dP}{dI} = N_p \frac{h\nu}{q}\frac{\alpha_m}{\alpha_{tot}}\left(\eta - \frac{\tau_2}{\tau_{eff}+\tau_2}\right) \tag{2.2.22}$$

$$R_{d,S>0}/R_{d,S=0} = \frac{1}{\eta}\frac{\tau_2}{\tau_{eff}+\tau_2} \tag{2.2.23}$$

2.2.5 Resonant tunneling

The electron transport in QCLs is controlled by resonant tunneling from to injector state into the upper state of the optical transition. Resonant tunneling injection assures an high injection efficiency and at the same time a very fast injection process, where the maximum current is only limited by the upper state lifetime in a properly designed injector. The first description of resonant tunneling goes back to Kazarinov and Suris, describing resonant tunneling between quantum well subbands [48]. Resonant tunneling injection has been described in a QCL framework [45]. The resonant tunneling of electrons from the injector state g to the upper state 3 of the laser transition in the tight-binding approximation leads to a current density

$$J = en_s \frac{2|\Omega|^2\tau_\perp}{1+\Delta^2\tau_\perp^2+4|\Omega|^2\tau_3\tau_\perp} \tag{2.2.24}$$

$$\hbar\Delta = E_g - E_3 = qd(F-F_r), \qquad d = |z_{gg} - z_{33}| \tag{2.2.25}$$

where e is the electronic charge, n_s the sheet density in the injector, $2\hbar|\Omega|$ is the energy splitting at resonance between the ground state of the injector (g) and the

$n = 3$ state, $\hbar\Delta$ is the energy detuning from resonance, d is the spatial separation between the centroids of the two electron probability distributions, F the average electric field applied over the distance d, and F_r is the electric field which brings the $n = 3$ and g states in resonance. The time constant τ_3 represents the lifetime of an electron in the $n = 3$ state and $\tau_\perp$ is the relaxation time for the momentum in the plane of the layers, responsible for the loss of phase between the states involved in resonant tunneling. When $\Delta = 0$, the maximum current density is

$$J_{max} = en_s \frac{2|\Omega|^2 \tau_\perp}{1 + 4|\Omega|^2 \tau_3 \tau_\perp} \tag{2.2.26}$$

The strong coupling regime for injection is if $4|\Omega|^2 \tau_3 \tau_\perp \gg 1$. The maximum current is controlled by the upper state lifetime τ_3 and reads $J_{max} = en_s/(2\tau_3)$. This is the regime that ensures fast electron injection in the upper state without being limited by the tunneling rate. QCL designs aim for the strong coupling regime. The upper state lifetime τ_3 and the coupling Ω are design parameters. The in plane dephasing time $\tau_\perp$ depends on the wavefunctions and growth parameters and cannot be calculated in a simple way. However it can be determined *a fortiori* by measuring the linewidth of the spontaneous emission.

If the applied voltage on the QCL is further increased after the injection resonance, a region of electrical instability is usually observed, paired with the break-down of the lasing regime. The electrical instability is called region of negative differential resistance (NDR) and occurs when the injector state is not aligned anymore with the upper state. NDR is used in resonant tunneling diodes [26, 7]. NDR in QCLs can be a small effect due to the presence of other subbands which create an important parallel channel for the electron transport.

Eq. 2.2.24 is a first-order formula for resonant tunneling. More recently a second-order mechanisms was developed [49]. It is found that resonant tunneling occurs with conservation of the energy rather than the wave vector, contrarily to the first-order case [50]. The second-order current between a pair of subbands coupled through a barrier in the particular case that both subbands share the same electronic temperature, and that a same uniform scattering potential is considered, is expressed as

$$J = d\frac{2e\Omega^2 2\gamma}{\Delta^2 + (2\gamma)^2}\left[\theta(\Delta)(n_2 - e^{-\beta\hbar|\Delta|}n_1) + \theta(-\Delta)(e^{-\beta\hbar|\Delta|}n_2 - n_1)\right] \quad (2.2.27)$$

where n_1 and n_2 are the electron densities in subbands 1 and 2, d is the difference between the two centroids of the wave functions $d = z_2 - z_1$, e is the elementary charge and γ is the broadening of the subbands, Δ is the detuning between the subband edges $\Delta = \epsilon_2 - \epsilon_1$, $\hbar\Omega$ the coupling energy through the barrier, $\theta(x)$ is the Heaviside function, and $\beta = 1/k_B T$, where k_B is the Boltzmann constant.

The current flow between two equally populated subbands is better described by the second order formalism. The first-order formalism predicts a zero net current for any detuning. In contrast, the second-order formalism predicts a dispersive shaped current density around the resonance. The current flow between two subbands, where one is empty, is no longer symmetric in respect to the detuning in the second-order formalism. However the first-order model is recovered as the thermal energy largely overcomes the detuning energy $k_B T \gg \hbar|\Delta|$.

2.3 Transport models

The electron transport in QCLs has been studied theoretically and several models have been developed to predict the transport, electron distribution, gain and the relative importance of different scattering mechanisms. Among the different models there are three main categories: Monte Carlo simulations, density matrix models and non-equilibrium Green's functions. Monte Carlo simulations constitute a semi-classical model for the transport in QCL [51, 52]. Scattering rates are computed between the different electronic states. However coherence phenomenons such as resonant tunneling and dephasing that are most important when describing the transport between two weakly coupled energy states are neglected. Simulations based on non-equilibrium Green's functions represent a full quantum mechanical approach [53]. However the complexity and large computational burden of this method limit its utility in obtaining an intuitive picture of the electron transport. The density matrix model is hybrid in the sense that it takes into account coherence phenomenons between weakly coupled quantum states, but uses otherwise semiclassical scattering rates [54, 55]. Simplified density matrix models have been developed for terahertz QCLs [46, 47, 56]. They are based on fitting parameters, but are useful for the optimization of a particular design.

A common result of all this models is that the electron temperature is significantly higher than the lattice temperature (~ 100 K), confirmed by experimental investigations on terahertz QCLs [57, 58]. The elastic scattering processes (alloy disorder, ionized impurity, interface roughness, electron-electron) contribute to the heating of the electron gas, whereas the inelastic scattering processes (optical and acoustical phonons) are responsible for energy dissipation to the lattice. Electron-electron scattering is responsible for the thermalization of electrons in subbands through a redistribution of the kinetic energy between the electrons [59]. The electron transport and population inversion is dominated by intersubband scattering. Intrasubband scattering affects the coherence and therefore broadening of the subbands. Absorption linewidths depend on both, intersubband and intrasubband scattering. It is usually an excellent approximation to assume that intrasubband and/or intrawell relaxtions are faster than intersubband and interwell ones [60].

2.4 Scattering

Fermi's golden rule is used to compute the total scattering rate $W = 1/\tau_i$ of a heterostructure eigenstate ψ_i into all the final states ψ_f induced by a scattering potential $V(\mathbf{r})$ [61]

$$W = \frac{2\pi}{\hbar} \sum_f |\langle \psi_i | V_{fi} | \psi_f \rangle|^2 \delta(\varepsilon_f - \varepsilon_i) \tag{2.4.28}$$

In the case of inelastic scattering, the energy conservation is adjusted accordingly. For example for the emission of a phonon it is replaced by $\delta(\varepsilon_f + \hbar\omega - \varepsilon_i)$. Using the notation of Ref. [60], the envelope conduction band states take the form

$$\psi_i(\mathbf{r}) = \frac{1}{\sqrt{S}} \exp(i\mathbf{k} \cdot \rho) \chi_i(z) \tag{2.4.29}$$

and have a corresponding energy

$$\varepsilon_i(k_i) = E_i + \frac{\hbar^2 k_i^2}{2m^*} \tag{2.4.30}$$

where S is the sample area, $\mathbf{k}_i$ is the in-plane component of the electron wave vector, E_i is the confinement energy of the ith subband, $\chi_i(z)$ the associated envelope function and m^* is the electron effective mass.

2.4.1 Optical phonon

The optical phonon (LO-phonon) interaction with an electron is mediated by the Fröhlich term. The scattering rate of an electron from subband i to subband f by emission of an LO-phonon at $T = 0$ K is given by [60]

$$W_{if}^{lo}(k_i) = \frac{m^* e^2 \omega_{LO}}{2\hbar^2 \varepsilon_0} \left(\frac{1}{\varepsilon_\infty} - \frac{1}{\varepsilon_r} \right) \int_0^{2\pi} d\theta \frac{I^{ij}(q_\perp)}{q_\perp} \tag{2.4.31}$$

where ε_∞ and ε_r are the high frequency and static relative dielectric constants, $\hbar\omega_{LO}$ is the LO-phonon energy (36 meV in GaAs) and $q_\perp$ is

$$q_\perp = \left(k_i^2 + k_f^2 - 2k_i k_f \cos\theta\right)^{1/2} \tag{2.4.32}$$

$$k_f^2 = k_i^2 + \frac{2m^*}{\hbar^2}(E_i - E_f - \hbar\omega_{LO}) \tag{2.4.33}$$

$I^{if}(q)$ is defined by

$$I^{if}(q) = \int dz \int dz' \chi_i(z)\chi_f(z) e^{-q|z-z'|} \chi_i(z')\chi_f(z') \tag{2.4.34}$$

which is equal to δ_{if} if $q = 0$ and which decays like q^{-1} at large q values. The transition rate for LO-phonon emission at non-zero temperatures is obtained by mulitying the zero-temperature result (Eq. 2.4.31) by (1+n), where $n = 1/(e^{\hbar\omega_{LO}/k_B T} - 1)$ is the thermal population of LO-phonons. Similarly, the transition rate for LO-phonon absorption is obtained by multiplying Eq. 2.4.31 by n and by changing $-\hbar\omega_{LO}$ into $\hbar\omega_{LO}$ in Eq. 2.4.33.

LO-phonon scattering is extremely fast, having typically sub-picosecond intrasubband and intersubband scattering lifetimes [60]. The depopulation of the lower laser subband in mid-infrared QCL designs is based on LO-phonon scattering [29]. In terahertz QCLs the subband spacing of the optical transition is lower than the LO-phonon energy. In this situation LO-phonon emission is still possible from hot electrons that have sufficient in-plane kinetic energy. An important mechanism for the degradation of the population inversion with temperature in terahertz QCLs is the thermal activation of LO-phonon emission from the upper laser subband [62]. LO-phonon emission has also been identified to be the main scattering mechanism at low lattice temperature ($T_l = 25$ K) that limits the upper state lifetime ($\tau^{lo} = 2.8$ ps) in terahertz QCLs based on the resonant phonon depopulation design [63].

2.4.2 Acoustical phonon

For acoustical phonons the scattering rate from subband i to f by emission of an acoustical phonon is given by [60]

$$W_{if}^{ac}(k_i) = \frac{D^2}{8\rho c_s^2 \pi^2 \hbar} \int_{-\infty}^{\infty} dq_\perp |f_{if}(q_\perp)|^2 \int_0^{2\pi} d\theta \tag{2.4.35}$$

$$\times \left[\int_0^\infty dk_f k_f \hbar\omega (n+1) \delta \left(E_i - E_f - \hbar\omega + \frac{\hbar^2 (k_i^2 - k_f^2)}{2m^*} \right) \right] \tag{2.4.36}$$

where

$$\omega = c_s \left(q_\perp^2 + |\mathbf{k}_i - \mathbf{k}_f|^2 \right)^{1/2} \tag{2.4.37}$$

$$f_{if}(q) = \int dz \chi_i(z) e^{-iqz} \chi_f(z) \tag{2.4.38}$$

D is the deformation potential for electrons, ρ the density, and c_s the longitudinal velocity of sound. The thermal population is given by $n = 1/(e^{\hbar\omega/k_B T} - 1)$. The scattering rate by absorption of an acoustical phonon is obtained by replacing $(n+1)$ by n and $-\hbar\omega$ by $\hbar\omega$ in the argument of the δ function.

Assuming high temperature compared to the typical acoustical phonon energy and linear phonon dispersion, the acoustical phonon scattering rate in the elastic scattering approximation is given by [64]

$$W_{if}^{ac} = k_B T \frac{m^* D^2}{\rho c_s^2 \hbar^3} \int dz [\chi_f(z) \chi_i(z)]^2 \tag{2.4.39}$$

where $c_l = \rho c_s^2$ is the longitudinal elastic constant.

The authors in Ref. [65] present calculations of the scattering lifetime for the emission of acoustical phonons in a GaAs quantum well for the $2 \rightarrow 1$ transition. Characteristic for acoustical phonon scattering is its ineffectiveness at low temperatures. The scattering lifetime increases with decreasing ΔE_{21} and strongly depends on the lattice temperature, particularly at small values of ΔE_{21}. At 4.2 K the computed $\tau_{21}^{ac} > 400$ ps and at 77 K $\tau_{21}^{ac} > 40$ ps for an intersubband separation in the terahertz ($E_{21} < 18$ meV). In Ref. [64] the transport in a terahertz QCL based on the resonant phonon depopulation design is studied with non-equilibrium Green's functions. Based on the elastic scattering approximation, it is found that acoustical phonon scattering contributes less than 1% to the total elastic scattering at a lattice temperature of $T = 100$ K.

2.4.3 Interface roughness

Monolayer fluctuations result in not atomically sharp interfaces. Roughness is modeled by a gaussian correlation function

$$\langle \Delta(\mathbf{r})\Delta(\mathbf{r}') \rangle = \Delta^2 \exp\left(-\frac{|\mathbf{r}-\mathbf{r}'|^2}{\Lambda^2} \right) \tag{2.4.40}$$

where Δ is the mean height of roughness and Λ is the correlation length. The scattering rate W_{if}^{ifr} is [66, 67]

$$W_{if}^{ifr}(k_i) = \frac{m^* \Delta^2 \Lambda^2 V_0^2}{\hbar^3} \sum_m |\chi_i^*(z_m)\chi_f(z_m)|^2 \int_0^\pi d\theta e^{-q_\perp^2 \Lambda^2/4} \tag{2.4.41}$$

where z_m is the position of the mth interface. The interfaces are assumed to be uncorrelated. The scattering vector $q_\perp$ is given by

$$q_\perp^2 = 2k_i^2 + \frac{2m^*(E_i - E_f)}{\hbar^2} - 2k_i\sqrt{k_i^2 + \frac{2m^*(E_i - E_f)}{\hbar^2}}cos\theta \qquad (2.4.42)$$

Interface roughness is equivalent to local fluctuations in the well width, it can be shown that $W_{if}^{ifr} \propto L^{-6}$ for a square well in the limit of the infinite barrier approximation.

The absorption linewidth of quantum well structures and QCL's at mid mid-infrared wavelengths (typically 100 - 350 meV) have been investigated theoretically and experimentally by several authors [68, 69, 70]. Interface roughness scattering is the dominant scattering mechanism for the absorption linewidth at cryogenic temperature and also at room temperature it is still dominating the LO-phonon scattering broadening.

The authors in Ref. [71] present an experimental and theoretical study of a GaAs/ AlGaAs mid-infrared QCL under a strong magnetic field. They show that at low temperature, the interface roughness scattering is the most efficient elastic relaxation mechanism between the radiative levels whenever LO-phonon emission is inhibited. Their interface roughness model is based on a bimodal roughness: substantial roughness on length scales greater than the exciton diameter and microroughness on length scales shorter than the exciton diameter. The latter is the interface roughness described above and leads to homogenous broadening. The roughness on greater length scales is motivated by "terrace like interface defects" [72, 73] and leads to inhomogeneous broadening that is modeled with a gaussian distribution [74]. Similar results were obtained in a recent report on a bound-to-continuum terahertz QCL based on $In_{0.53}Ga_{0.47}As/In_{0.48}Al_{0.52}As$ [75]. The upper state lifetime of the radiative transition is computed by considering alloy, interface roughness and LO-phonon scattering. Comparison with experiment suggests that the elastic processes limit the lifetime at low temperatures, while the inelastic LO-phonon scattering becomes the dominant scattering mechanism at elevated temperatures. The scattering rate of the upper lasing level by interface roughness scattering is $\tau^{ifr} = 15.2$ ps and by alloy disorder scattering $\tau^{ad} = 17.8$ ps. Terahertz QCLs based on intrawell transitions and on bound-to-continuum transitions have been studied in a strong magnetic field [76]. Extremely low values for the threshold current density are observed for the QCL's based on an intrawell transition. Structures based on bound-to-continuum transitions, where the wave function is delocalized, show only a moderate reduction of the threshold current density, which is attributed to interface roughness scattering.

2.4.4 Alloy disorder

In ternary layers composed of $A_xB_{1-x}C$, such as $Al_xGa_{1-x}As$, electrons are scattered by conduction band disorder. The scattering rate W_{if}^{ad} is [67]

$$W_{if}^{ad} = \frac{m^* a^3 (\delta E_c)^2 x(1-x)}{\hbar^3} \int_{alloy} dz \left|\chi_f^*(z)\chi_i(z)\right|^2 \tag{2.4.43}$$

where a is the lattice constant and δE_c is the difference in the conduction band minima of crystals AC and BC (AlAs and GaAs in the present case). W_{if}^{ad} is independent on the in-plane wave vector $\mathbf{k}_i$. Alloy disorder scattering can be regarded as a kind of roughness scattering. If one substitutes $V_0 = \delta E_c$, $\Delta^2 = a^2 x(1-x)/4$, and $\Lambda^2 = a^2/2\pi$, one can recognize that the alloy disorder scattering is expressed as the sum of the roughness scattering rates due to the alloy layer at position z.

For InGaAs/AlInAs based mid-infrared QCL's alloy scattering has been identified to be the dominant elastic relaxation scattering mechanism that has a weight comparable to LO-phonon scattering [77]. Similar, in a terahertz QCL based on the InGaAs/AlInAs material system, alloy disorder scattering was shown to have a weight comparable to interface roughness scattering, and is one of the main scattering mechanisms for the upper state at low temperature [75]. Alloy disorder scattering is expected to be negligible for the GaAs/AlGaAs material system, since the well material is a binary compound. For this reason alloy disorder scattering is usually not included in Monte Carlo simulations and non-equilibrium Green function models of GaAs/AlGaAs based terahertz QCL's.

2.4.5 Ionized impurity

When dopant donors of Si are ionized, electrons supplied to quantum wells suffer from scattering by the Coulomb potential of the donors. The scattering rate is [67, 78]

$$W_{if}^{ion}(k_i) = \frac{m^*}{\pi\hbar^3}\int dZ N(Z)\int_0^{\pi} d\theta\left|V_{ij}^s(q_\perp, Z)\right|^2 \tag{2.4.44}$$

where $N(Z)$ is the volume impurity concentration at position Z, V^s_{ij} is the screened impurity scattering matrix element and the scattering vector $q_\perp$ is given by

$$q_\perp^2 = 2k_i^2 + \frac{2m^*(E_i - E_f)}{\hbar^2} - 2k_i\sqrt{k_i^2 + \frac{2m^*(E_i - E_f)}{\hbar^2}} cos\theta \quad (2.4.45)$$

The unscreened impurity scattering matrix element is

$$V_{ij}(q, Z) = \frac{-e^2}{2\varepsilon_0\varepsilon_r q}\int dz\, \chi_i^*(z)\chi_j(z)e^{-q|z-Z|} \quad (2.4.46)$$

Screening of the impurity potential by the conduction band electrons is an important effect. Screening will be discussed together with screening of the electron-electron interaction.

2.4.6 Electron-electron

The electron-electron scattering rate of an electron with wave vector $\mathbf{k}_i$ and spin σ in subband i and a second electron with wave vector $\mathbf{k}_j$ and opposite spin in subband j into the final states with wave vectors $\mathbf{k}_f$ and $\mathbf{k}_g$ and subband indices f and g is given by [65]

$$W^{ee}_{i,j\rightarrow f,g}(\mathbf{k}_i, \sigma) = \frac{4m^*}{\pi\hbar^3}\int d\mathbf{k}_j \int_0^{2\pi} d\theta \left|V^s_{ifjg}(q_\perp)\right|^2 f_j(\mathbf{k}_j)[1 - f_f(\mathbf{k}_f)][1 - f_g(\mathbf{k}_g)] \quad (2.4.47)$$

$f_i(\mathbf{k}_i)$ is the carrier distribution function that gives the occupation probability of state $\mathbf{k}_i$ in subband i and V^s_{ifjg} is the screened Coulomb scattering matrix element.

The unscreened coulomb interaction is given by

$$V_{ijkl}(q) = \frac{e^2}{2\varepsilon_0\varepsilon_r}\frac{F_{ijkl}(q)}{q} \tag{2.4.48}$$

The form factor is defined by

$$F_{ijkl}(q) = \int dz \int dz' \chi_i^*(z)\chi_j(z)e^{-q|z-z'|}\chi_k^*(z')\chi_l(z') \tag{2.4.49}$$

The scattering vector $\mathbf{q}_\perp$, and the vector $\mathbf{k}_{ij}$ are defined by

$$\mathbf{q}_\perp = \mathbf{k}_i - \mathbf{k}_f \qquad \mathbf{k}_{ij} = \mathbf{k}_j - \mathbf{k}_i \tag{2.4.50}$$

The magnitude of the scattering vector $q_\perp$ is expressed as

$$q_\perp(k_{ij},\theta) = \left[2k_{ij}^2 + \Delta k_0^2 - 2k_{ij}\sqrt{k_{ij}^2 + \Delta k_0^2}cos\theta\right]^{1/2}/2 \tag{2.4.51}$$

$$\Delta k_0^2 = \frac{4m^*}{\hbar^2}(E_i + E_j - E_f - E_g) \tag{2.4.52}$$

where θ is the angle between the relative wave vectors $\mathbf{k}_{ij}$ and $\mathbf{k}_{fg}$.

2.4.7 Screening models

The screened impurity scattering matrix element V_{ij}^s due to screening of the electrons in the subband 0 is computed in Ref. [78]. The screening is included by considering the diagram shown in Fig. 2.4. If the polarizability due to virtual intersubband transitions is neglected, the screened impurity scattering matrix element is given by

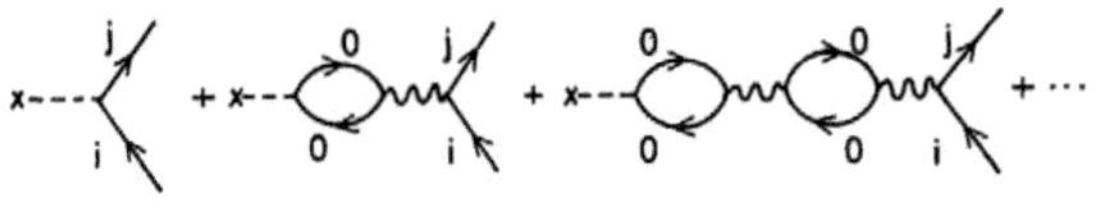

Figure 2.4: (Figure and description reprinted from [78]). A diagrammatic representation of matrix elements of impurity scatterings. The cross denotes an impurity and wavy lines represent the Coulomb interaction. The screening charge is assumed to arise only from electrons in the lowest subband 0 and inter-subband excitations are neglected.

$$V_{ij}^{s}(q,Z) = V_{ij}(q,Z) + V_{00}(q,Z)[\tilde{\varepsilon}(q)^{-1} - 1]\frac{F_{00ij}(q)}{F_{0000}(q)} \tag{2.4.53}$$

where the static dielectric function $\tilde{\varepsilon}(q)$ is given by

$$\tilde{\varepsilon}(q) = 1 + \Pi(q)V_{0000}(q) \tag{2.4.54}$$

This formalism can be extended to many-subband cases [66]. Then the matrix elements of the dynamically screened interaction are coupled to each other in a system of simultaneous linear equations.

The screened electron-electron scattering matrix elements are obtained by the equation [79]

$$V_{ijkl}^{s}(q) = V_{ijkl}(q) + \sum_{m,n} V_{ijnm}(q)\Pi_{mn}(q)V_{mnkl}^{s}(q) \tag{2.4.55}$$

and similar, the screened impurity scattering matrix elements are obtained by [80]

$$V_{ij}^{s}(q) = V_{ij}(q) + \sum_{m,n} V_{ijnm}(q)\Pi_{mn}(q)V_{mn}^{s}(q) \tag{2.4.56}$$

The polarization function Π_{mn} is the important quantity where most of the approximations are usually done. Different screening models will be discussed in the following. The random phase approximation is the most basic approximation and represents a starting point for different screening models with further approximations.

Random phase approximation

The polarization function reads in the random phase approximation [79, 81]

$$\Pi_{mn}(\mathbf{q}) = \lim_{\delta \to 0} \frac{1}{A} \sum_{\mathbf{k}} \frac{f_{m,\mathbf{k}+\mathbf{q}} - f_{n,\mathbf{k}}}{E_{m,\mathbf{k}+\mathbf{q}} - E_{n,\mathbf{k}} - i\delta} \tag{2.4.57}$$

where A is the in-plane surface and $f_{m,\mathbf{k}}$ is the occupation of state $\mathbf{k}$ in subband m.

Static long-wavelength limit

The static long-wavelength limit is obtained when taking the limit $(q \longrightarrow 0)$ in Eq. 2.4.57. The polarization function becomes

$$\Pi_{ii} = -\frac{m^*}{\pi\hbar^2} f_{i,\mathbf{k}=0} \qquad \Pi_{ij} = \frac{n_i - n_j}{E_i - E_j} \quad \text{if } i \neq j \tag{2.4.58}$$

where n_i is the sheet carrier density in the ith subband. The static long-wavelength limit model is often referred as the multisubband screening model. In Ref. [63] a multisubband screening model has been implemented for electron-electron and electron-impurity scattering in a Monte Carlo simulation of a terahertz QCL based on the resonant phonon depopulation design. The inclusion of impurity scattering eliminates the problem of underestimation of the current densities of previous simulations without impurity scattering. Furthermore it results in an increase of the electron temperature. The authors find that for intersubband transport electron-impurity scattering usually dominates over electron-electron scattering close to the doping layer. In particular they simulate between the radiative levels $\tau_{54}^{ee} = 49$ ps and $\tau_{54}^{ion} = 23$ ps.

Two subband screening model

If only two subbands are considered for screening, Eq. 2.4.55 can be solved analytically, giving simple expressions for the screened Coulomb matrix elements. For an intersubband transitions in a symmetric potential, where both electrons change the subband, the screened Coulomb matrix element can be expressed as [79]

$$V_{1212}^{s}(q,\omega) = \frac{q}{q+\kappa_{12}(q)} V_{1212}(q) \tag{2.4.59}$$

where $(V_{1212}^{s} = V_{2121}^{s} = V_{1221}^{s} = V_{2112}^{s})$, and the screening wave number $\kappa_{12}(q)$ is

$$\kappa_{12}(q) = \frac{e^2(n_1 - n_2)}{\varepsilon_0 \varepsilon_r (E_2 - E_1)} F_{1212}(q) \tag{2.4.60}$$

Note that the form factor F_{1212} tends linear to zero as $q \to 0$, therefore, V_{1212} and V_{1212}^{s} are finite for $q = 0$. Similar relations exist for the two band screening of V_{1111}^{s} and V_{2222}^{s}.

Single subband screening model

If only a single subband is considered, Eq. 2.4.55 reduces to

$$V_{1111}^{s}(q,\omega) = \frac{q}{q+\kappa_1(q)} V_{1111}(q) \tag{2.4.61}$$

where the quasi-2D screening wave number is

$$\kappa_1(q) = \frac{m^* e^2}{2\pi\varepsilon_0\varepsilon_r\hbar^2} F_{1111}(q) f_{1,\mathbf{k}=0} \tag{2.4.62}$$

The authors in Ref. [79] point out that the single subband dielectric screening function is sometimes used in literature to model the screening for intersubband transitions. The screened potential differs qualitatively for small q values. While it tends to a finite value using κ_{12}, the use of κ_1 leads to a screened potential that goes to zero for $q \to 0$, leading to a gross underestimation of the scattering rate.

In Ref. [80] different screening models for electron-electron and electron-impurity scattering are compared in a Monte Carlo simulation of a quantum cascade structure. The authors compare the multisubband screening model to two different single subband screening models. The single subband screening models are based on a constant screening wave vector obtained from Eq. 2.4.62 by $q_s = \kappa_1(q \to 0)$. This screening wave vector is used to compute the screened potentials. In the first model they assume that all the carriers are in the lowest subband and form a quasi Fermi distribution. The second model takes only into account the electron distribution in the initial subband. Their simulations show that one single subband model overestimates screening, the other underestimates it compared to the multisubband screening model. They propose a modified single subband screening model that takes into account screening form all the temperature different subbands, lumped in one screening length:

$$q_s = \frac{m^* e^2}{2\pi \varepsilon_0 \varepsilon_r \hbar^2} \sum_i f_{i,\mathbf{k}=0} \tag{2.4.63}$$

where i sums over the subbands in one period. In Ref. [82] the authors performed Monte Carlo simulations on a bound-to-continuum terahertz QCL. They compare a full random phase approximation screening model to the modified single subband model (Eq. 2.4.63) for the implementation of electron-electron and electron-impurity scattering. The peak gain computed with the modified single subband model differs by approximately 25% compared to the full random phase approximation. In the modified single subband screening model the intersubband scattering rate is underestimated due to an overestimation of the screening. The occupation of the miniband levels are quite insensitive to the chosen screening model. This is consistent with the fact that the average inter and intrasubband scattering time of an electron strongly depend on the chosen screening model, but is changed by a similar factor for all subbands.

Isotropic screening model

A simple isotropic screening model in the context of impurity scattering has been considered in Ref. [64, 81]. The screened impurity matrix element is obtained by the replacement $q \to \sqrt{q^2 + q_s^2}$ on the right-hand side of Eq. 2.4.46. q_s is the inverse screening length, which in the Debeye approximation (valid if $k_BT > \mu$) becomes

$$q_s^2 = \frac{e^2 N}{\varepsilon_r \varepsilon_0 k_B T} \tag{2.4.64}$$

where N is the average electron volume density in the structure and μ is the chemical potential of the corresponding three dimensional electron gas. In Ref. [64] the authors investigate the evolution of gain in a phonon depopulation based terahertz QCL, using a non-equilibrium Green's function method. They included LO-phonon, impurity and interface roughness scattering in their model. They argue that the increased broadening, primarily due to increased impurity scattering and not diminishing population inversion is the main reason for the reduction of peak gain with temperature. Interface roughness scattering of the upper and lower state is small compared to impurity scattering and the same holds for the LO-phonon emission from the lower laser state. The screening of ionized impurity scattering uses the Debeye approximation where the electrons contributing to screening are assumed to be in a thermal equilibrium and obey Boltzmann statistics. The inverse screening length q_s is temperature dependent. At low temperature, the Thomas-Fermi approximation is used to compute the screening length. The full random phase approximation, the multi-subband screening model and an isotropic screening model with a temperature dependent screening length in the Debeye approximation for the calculation of impurity scattering are compared by the same authors [81]. The full random phase approximation and the multi-subband model lead to almost identical results. If the screening length is of the order of the laser period or longer, the isotropic bulk screening model is an excellent approximation to the full random phase model. In this situation, which is common for typical terahertz QCLs, the distribution of electrons to different subbands is much less important than the average subband temperature.

Chapter 3

Experimental methods

3.1 Sample fabrication

3.1.1 Growth

The samples are grown by molecular beam epitaxy on GaAs substrates. The growth begins with a 200 nm thick $Al_{0.5}Ga_{0.5}As$ or a 300 nm thick $Al_{0.3}Ga_{0.7}As$ etch stop layer, followed by a 60 nm thick highly doped bottom contact layer (GaAs, Si-doped, $n = 3 \times 10^{18}$ cm^{-3}), then typically $60 - 120$ repetitions of the active region are deposited and finished by a highly doped top contact (GaAs, Si-doped, 60 nm with $n = 3.8 \times 10^{18}$ cm^{-3} and InGaAs, 20 nm with $n = 2 \times 10^{19}$ cm^{-3}, Si-doped). The bottom contacts of the layers V303, N891, N892, N899, N908 differ in their thickness ($400 - 700$ nm) and doping level (1×10^{18} cm^{-3} to 2×10^{18} cm^{-3}).

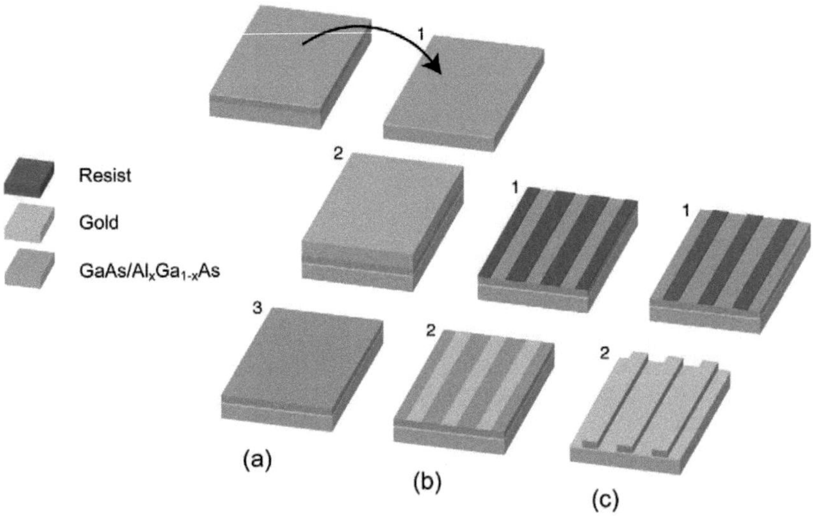

Figure 3.1: Double metal laser fabrication. (a): Flip-chip process step of the active region. The sample and a hosting substrate are coated with gold, followed by waferbonding and substrate removal. (b): Top metallization of the double metal waveguides. The photolithography in negative resist is followed by a gold coating and lift-off. (c): Etching of the laser ridges. The etch mask is transferred into a positive resist and ridges are defined by wet etching.

3.1.2 Double metal ridge lasers

The double metal waveguide requires a flip-chip process, since the active region has to be sandwiched between two metallic layers [see Fig. 3.1(a)]. Samples with a size of typically 1 x 1 cm are cleaved. A sample with the grown epitaxial-layers, and a doped GaAs hosting substrate are coated with Ti/Au (5/500 nm). Both samples are aligned and pressed manually together with their gold surfaces touching. Then thermo-compression wafer-bonding [83] is carried out by heating of the sample to 300°C and applying a pressure of 4.5 MPa during 45 minutes. After cooling the wafer-bonded sample down to room temperature, the original substrate is removed first by mechanical lapping, the remaining 50 μm are removed by a chemical etchant based on citric acid (Table A.1 on page 207). The $Al_xGa_{1-x}As$ layer acts as an etch stop layer for this etching solution when $x \geq 0.3$. The etch stop layer is removed by a non-selective phosphoric etchant for $x = 0.3$ or with highly selective pure hydrofluoric (HF) acid for the samples with $x = 0.5$. The active region is now up-side down, with a buried gold layer, on the hosting substrate.

The top metallization of the double metal waveguide is performed in the next process step [Fig. 3.1(b)]. A negative resist, such as the MAN1420, is used to define ridges by photolithography. The sample is then coated with Ti/Au (5/250 nm), followed by a lift-off in aceton.

Laser ridges are defined by wet etching [Fig. 3.1(c)]. The metallic defined ridges are covered with a positive resist mask (AZ1518). Etching is performed in a sulfuric acid etching solution (Table A.1). The sample is immersed parallel to the surface in the etchant, hold laterally by tweezers, and moved slowly back and forth, with a frequency of about 1 Hz. The sulfuric acid etchant is anisotropic. Depending on the orientation of the ridges in respect to the crystallographic orientation of GaAs, different etch profiles are obtained [84, 85]. Low frequency terahertz QCLs with the inward sloping profile have shown better performance than with the outward sloping profile. A possible reason is the more uniform current density in the active region or lower waveguide losses. After etching, the resist is removed with aceton. The process is finished by a not mandatory backside lapping to thin down the substrate to about 200 μm for improved thermal heat conductivity, followed by a Ti/Au (10/200 nm) coating, required for mounting of the sample with an indium solder.

3.1.3 LC laser

As for the double metal laser process, the fabrication starts by metallization and bonding of the grown active region on a carrier substrate by thermo-compression wafer-bonding. The etch stop layer is a 200 nm thick $Al_{0.5}Ga_{0.5}As$ layer. It is not removed, since it serves as electrical isolation layer for the bonding pad [Fig. 3.2(a)]. In a first process step the $Al_{0.5}Ga_{0.5}As$ layer is etched in the areas where the LC laser resonators will be defined [Fig. 3.2(b)]. A positive resist (AZ1518) is used to mask the un-etched areas. A phosphoric acid based etchant is used (Table A.1). For an etch depth of 200 nm the etchant acts isotropic, the edges have an outward slope, required for the following metallization. Hydrofluoric acid could be used in principle for this etch step, with the advantage of good selectivity to GaAs, but with the drawback that it diffuses through the resist and the edges have not a suitable profile for continuous metallization.

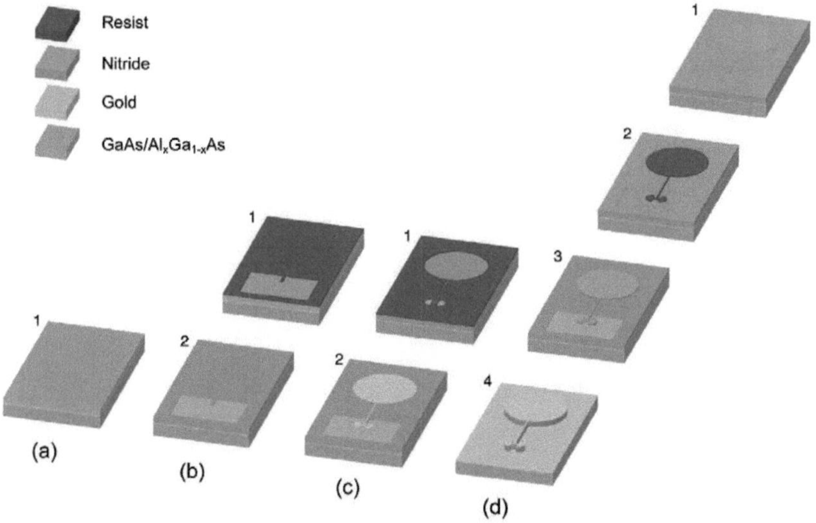

Figure 3.2: LC laser fabrication. (a): Starting point is a waferbonded sample. The $Al_{0.5}Ga_{0.5}As$ layer is not removed. (b): Partial etching of the $Al_{0.5}Ga_{0.5}As$ layer for electrical pumping of the resonator. A positive resist is used for the definition of the etching mask. (c): Top gold metallization of the resonator and pad. An image reversal resist is used for the transfer of the mask by photolithography, followed by gold coating and lift-off. (d): Dry etching of the LC laser. Nitride deposition is followed by a positive resist photolithography, and dry etching of the nitride. The nitride mask is used for the dry etching step of the active region and afterwards removed with diluted HF.

In the next step, the top metallic layer is defined [Fig. 3.2(c)]. The negative resist MAN1420 can be used, but the image reversal resist AZ5214 leads to better results. Since very small features (down to 2 μm) have to be transferred into the resist, it is crucial that the resist is perfectly flat. The resist is naturally thicker at the edge of the sample, causing an edge bead problem that makes proper exposure during contact lithography difficult. The resist close to the edge of the sample can be removed in a first exposure–development step when working with the image reversal resist; a procedure that is impossible with negative resist. The photolithographic step is followed by coating, Ti/Au (5/350 nm), and lift-off.

The last process step consists in dry etching of the active region to define the LC resonator and pad structure [Fig. 3.2(d)]. A nitride mask is used for dry etching, therefore nitride is deposited on the whole sample in a PECVD at 120°C (Sec. A.3). A resist mask (AZ1518) that covers exactly the gold shape below the nitride is defined by photolithography. The resist mask is nominally 500 nm wider than the gold shape. Sub-micrometer precision alignment is absolutely crucial. The resist pattern is transfered to the nitride by dry etching in an RIE. The resist mask is removed in aceton, followed by dry etching of the active region in an ICP (Sec. A.3). Timing is important during this etching step, since ideally the active region is etched down to the buried gold layer, but stopped immediately before the latter is reached. Bad timing leads to sputter-etching of the gold layer, that covers the side walls of the LC laser with sputtered gold, resulting in high optical losses and electrical short cuts. After dry etching the nitride mask is removed in diluted hydrofluoric acid. Since hydrofluoric acid is an etchant for titanium and adhesion problems of the gold layer during wire bonding on the pad are observed, the nitride

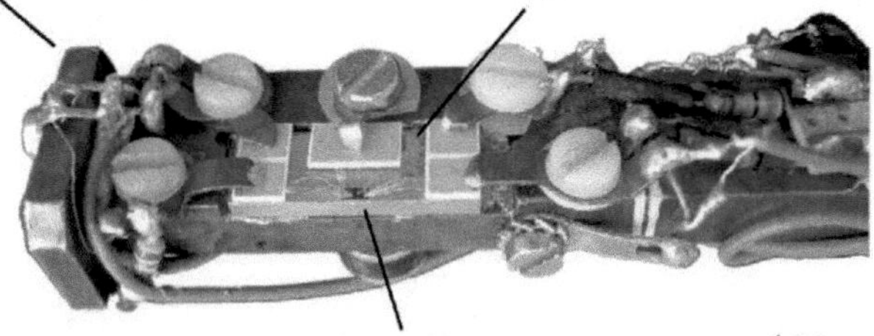

Figure 3.3: A sample is fixed on the cold finger of the helium flow cryostat. The sample has several contacted laser bars, the bonding wire between the laser bars and pads are visible.

mask might possibly be removed in a safer way by an RIE etching of the nitride. Backside coating of the sample with Ti/Au (10/200 nm) finishes the process. Further details on the processing parameters, especially the etching solutions, resists and plasma processing are found in the Appendix A.

3.2 Characterization

After processing, samples are cleaved, indium soldered on copper mounts and measured on the cold finger of a temperature-controlled helium flow cryostat, shown in Fig. 3.3. The finger of the cryostat is inserted in a vacuum shield that is equipped with windows for the terahertz.

3.2.1 Spectral characterization

A vacuum Fourier transform infrared (FTIR) spectrometer is used for spectral characterization of the samples. The FTIR is based on the Michelson-Moreley interferometer. The resolution is given by the optical path length difference between two interfering light beams. The interference signal is measured with a bolometer (from IRLabs) that has a noise equivalent power in the range of $pW/\sqrt{Hz}$ range. Measurements are carried out in cw mode or pulsed mode laser operation using either the rapid scan or step scan acquisition mode of the FTIR. In the step scan mode a lock-in amplifier is used for a phase sensitive detection. Low resolution spectra are measured with a custom vacuum FTIR and high resolution spectra are measured with a Bruker Vertex 80V that has a maximum resolution of 0.075 cm^{-1}. The custom vacuum FTIR and bolometer are described in reference [86].

3.2.2 Laser power characterization

The power and voltage versus current characterization of double metal lasers is carried with an absolute terahertz power meter from Thomas Keating Instruments. The power meter is based on an opto-acoustical detection of the power. The power meter head, shown in Fig. 3.4, comprises a closed air-filled cell formed by two closely-spaced parallel windows with a thin metal film in the gap between them. A known fraction (close to 50%) of the power of the incident beam is absorbed in the metal film, the remainder being partly transmitted through the cell and partly reflected from it. The beam is to be 100% amplitude-modulated either at its source or by a chopper placed in the path of the beam. The modulation frequency is to be in the range of 10 Hz to 50 Hz. The absorption of the power in the film produces modulated variations in the temperature of the film and of the layers of air in contact with the film. In turn this produces a modulation of the pressure in the cell which is detected by a pressure-transducer, measured by

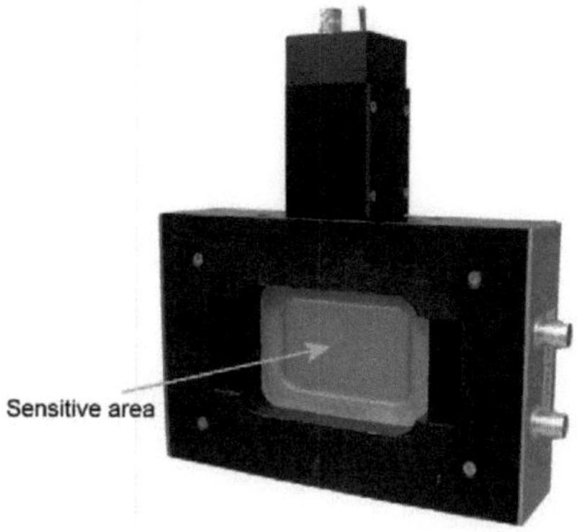

Figure 3.4: Absolute power meter head from Thomas Keating Instruments.

the lock-in amplifier. The modulated pressure change is closely proportional to the total absorbed power; it depends negligibly on the distribution of the absorbed power over the film. To calibrate the power meter, a measured amount of modulated ohmic power is generated in the film, by passing a modulated current through it via electrodes which contact opposite edges of the film. A look-up table in the software gives the window-loss as a function of frequency and fractional film absorption. The noise equivalent power is typically $5\mu W/\sqrt{Hz}$. The power meter is specified for frequencies from 30 GHz to 3 THz. The sensitive area is greater than 30 mm in diameter.

3.2.3 LC laser power characterization

Since the power of the LC laser is in the pW to nW range, a bolometer is used for the measurement. The specified voltage/power response of the bolometer is used to convert the measured voltage into power units. Measurements are carried out in

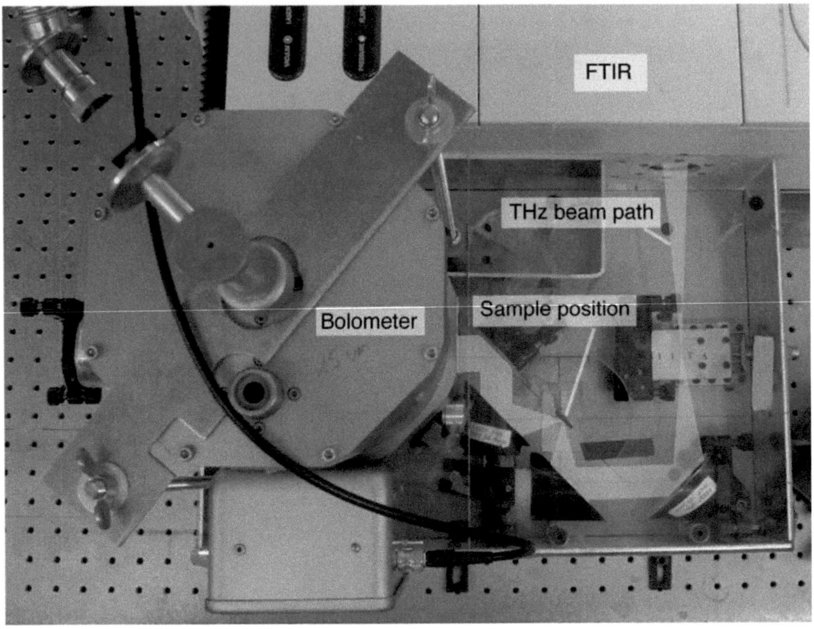

Figure 3.5: Picture of the reflection setup. The THz beam path is indicated in yellow. A cryostat (not shown) is used to place the sample in the focal point of the parabolic mirrors. The box containing the optics is purged with nitrogen to avoid the absorption of terahertz radiation by water vapor.

the custom FTIR setup, with the mirror at the zero path difference position.

For the measurements of the LC laser in the magnetic field, a helium cryostat with a superconducting coil (from Janis) is used. Measurements with magnetic fields up to 12 T or even 14 T when employing the lambda plate refrigerator are possible. The insert probe is equipped by an InSb cyclotron resonance detector that is used for the power measurement in the magnetic field. For more details on the setup, the reader is referred to the detailed description in reference [87].

3.2.4 Reflection measurement

The measurement of the frequency dependent reflectivity is carried out using the Bruker Vertex 80V spectrometer with a broad-band mid-infrared glow bar source and a bolometer detector. The interference beam of the glow bar is coupled out of the FTIR, focussed on a sample with 2 inch off axis parabolic mirrors (#1 optics) and then measured with a bolometer. The incidence on the sample is at an angle of 45 degrees. A picture of the setup is shown in Fig. 3.5.

Chapter 4

Low frequency terahertz QCLs

4.1 Terahertz active region design

In a QCL, the population inversion is maximized by achieving the largest ratio of upper to lower lifetime τ_{32}/τ_2, a large injection efficiency in the upper state η, and a long upper state lifetime τ_3. Three main classes of terahertz designs are distinguished and are shown in Fig. 4.1. In all schemes carrier injection is based on resonant tunneling from the lowest injector state into the upper state of the laser transition. The variety of different designs is impressive; at least ten active region designs have been published so far and will be discussed briefly: chirped super lattice, bound-to-continuum, bound-to-continuum with LO-phonon extraction, resonant phonon depopulation, step-well, two well, interlaced, bound-to-continuum with split injector, single quantum well, big well.

The active region of the first terahertz QCL is based on a chirped superlattice active region [32], in which population inversion between the two states at the edge of the minigap between two minibands is obtained by a phase space argument shown in

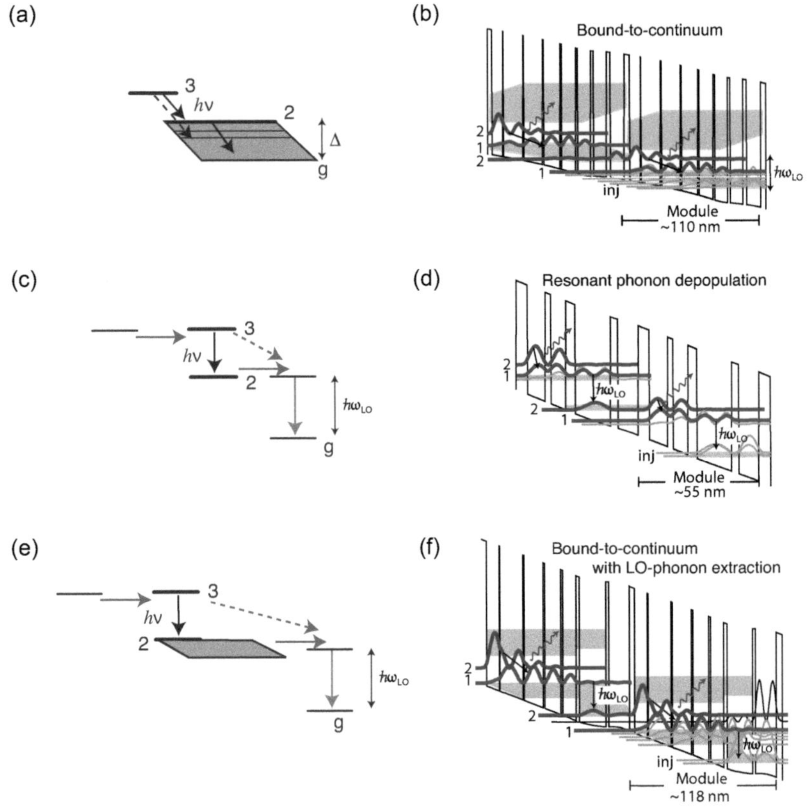

Figure 4.1: (a): Population inversion is achieved due to a phase space argument with a miniband, (b): example of the bound-to-continuum design. (c): Population inversion is based on resonant tunneling followed by LO-phonon scattering, (d): example of the resonant phonon depopulation design. (e): Population inversion by combining the miniband approach with an LO-phonon scattering extraction, (f): example of the bound-to-continuum with LO-phonon extraction design.

Fig. 4.1(a). The vertical radiative transition is characterized by a large dipole, but a relatively short upper state lifetime. The highest operation temperature achieved with this design is 75 K [88]. The main limitations in the performance, namely the poor slope effciency and the low maximum operating temperature, are attributed to thermal backfilling of the lower miniband and a weak population inversion. In the bound-to-continuum design [89], the population inversion is achieved as well by a phase space argument, but in contrast to the chirped super lattice design, the isolated upper state in the minigap is created by a thin well adjacent to the injection barrier, shown in Fig. 4.1(b). The diagonal transition between the upper state and the miniband reduces the spatial overlap, enhancing the upper state lifetime and the population inversion. The bound-to-continuum approach yields devices with a very high slope efficiency and power at temperatures up to 40 K [90, 91]. However, the maximum operating temperature of these devices remains limited to about 100 K. For the miniband and bound-to-continuum approach to succeed, a miniband width much larger than the broadening of the individual levels $\Delta \gg \Gamma$ and larger than the thermal energy $\Delta > k_B T$ should be maintained. Since the miniband width is only in the order of 15 meV, these conditions are only satisfied at low temperature [92].

Coupling of the lower state of the laser transition by resonant tunneling to a very short-lived upper state of a nearby well allows a significant reduction of the lower state lifetime while preserving a long upper state lifetime [see Fig. 4.1(c)]. This idea is the basis of the resonant phonon depopulation design [93], shown in Fig. 4.1(d). The design has seen many variations [94, 95] and the current version has demonstrated the highest operating temperature of 186 K achieved by a terahertz QCL

[20]. The low lifetime of the upper level of the well following the active region is usually achieved by spacing the levels resonantly with the LO-phonon energy. The recently demonstrated step-well [96] aims for the reduction of the leakage current prior to alignment in the phonon depopulation design. A grading of the composition of the quantum well where the optical transition is located, efficiently reduces the coupling of the injector to other states prior to alignment. A low threshold current density of only 110 A/cm^2 for a phonon depopulation design is obtained with a maximal operation temperature of 123 K. In the so called two-well design [97] the sub-picosecond scattering time of the electrons in the lower state by LO-phonon scattering is exploited to obtain a robust population inversion, which does not rely on a resonant tunneling extraction of the carriers as in the resonant phonon depopulation based structures. Only three states are involved in the transport in this simple design, resulting in a photon-phonon-resonant tunneling transport scheme. The highest operation temperature achieved with this design is 125 K [98].

A third class of designs aims at combining the advantages of both approaches, illustrated in Fig. 4.1(e). The active stage is based on a bound-to-continuum transition. The miniband is coupled to an extractor quantum well that enables fast LO-phonon scattering [99], shown in Fig. 4.1(f). This architecture has the advantage of reducing the direct coupling between the upper state and the extractor well since they are physically separated by the length of the miniband region. A maximum operation temperature of 160 K is reported for the bound-to-continuum design with LO-phonon extraction, by optimization of the miniband [92]. A similar strategy is followed by the so-called interlaced design, where the chirped superlattice design is combined with an LO-phonon extraction stage [100]. The bound-to-continuum design with split injector [101] has been developed for low frequency operation and resembles to a scaled version of to the bound-to-continuum design with LO-phonon extraction, but the extraction stage is far from being on LO-phonon resonance.

Single quantum well active regions where population inversion is established between two successive levels, leading to a vertical intra-well optical transition, have been explored. Electrons are extracted from the lower state by resonant tunneling into a miniband. The single quantum well design is based on a population inversion between the first excited state and the ground state of the well [56]. In the so-called big well design a population inversion is established between the second and first excited state [102] or even between the third and second excited state [103]. The lifetimes of the higher excited states in the quantum well are very short. Scattering rates are efficiently quenched by applying a strong magnetic field perpendicular to the quantum wells, resulting in a three dimensional confinement of the electrons. Record threshold current densities below 1 A/cm^2 are observed [104]. The maximum operation temperature of a big well design is 75 K [105].

4.1.1 Active regions below 2 THz

The extension of the operation range of QCL down to 1 THz is an important milestone for filling the terahertz gap with coherent sources. Different terahertz QCL designs have been explored for operation < 2 THz.

The bound-to-continuum based design is scaled down to 2 THz, with high peak powers up to 50 mW [91]. The miniband energy dispersion exceeds the photon energy. The advantages of this design are a good injection efficiency in the isolated upper state via resonant tunneling and a fast depopulation of the lower state via intra-miniband elastic scattering. The major challenge is to reduce the intersubband absorption and the leakage current prior to the alignment. Resonant cross absorption at the photon energy is reduced by spatially chirping the wavefunctions. However a very high absorption resonance at low energy is intrinsically related to the injector miniband and its high energy tail could constitute a fundamental issue for this design at even lower laser frequencies. No operation below 2 THz based on this design is reported.

The resonant phonon depopulation scheme with a one-well injector [106] efficiently reduces the intersubband absorption at low frequency, due to the lack of additional subbands in the one-well injector. Laser operation up to 110 K in pulsed mode is achieved at 1.9 THz. The major drawback of this design is the strong parasitic coupling of the injector state to the lower state of the laser transition prior to the alignment bias. To reduce this parasitic coupling thicker injector and collector barriers and a spatially diagonal radiative transition are employed. Marginal laser operation at 1.6 and 1.45 THz are achieved with this design, but the strong parasitic current channel to the lower state affects negatively the dynamic range, performance and further scaling possibilities [107].

The big well design is scaled successfully from 3.6 THz to 0.95 THz [105]. The requirement of a strong magnetic field being applied to the laser may limit potential applications of lasers based on this design.

4.2 Design of low frequency terahertz QCLs

4.2.1 Challenges below 2 THz

The challenges for terahertz QCL's below 2 THz arise from the very low photon energy $E_{ph} < 8$ meV. The energy spacing between the radiative states becomes comparable to the broadening of the states and to the energy coupling between states. The main challenges are listed below.

1. *Optical losses.* Low loss waveguides are of crucial importance. In a Drude bulk approximation the free carrier absorption has a λ^2 wavelength dependence and may introduce severe difficulties at long wavelengths, respectively low frequencies.

2. *Lifetimes.* The low energy separation between the radiative states makes lifetime engineering in the active region a difficult task. An efficient extraction of electrons from the lower radiative state is required with minimal reduction of the upper state lifetime. The scattering time from the upper to the lower state of the optical transition must be significantly larger than the lower state lifetime.

3. *Injection efficiency.* A necessary requirement for a sufficient population inversion is a large injection efficiency from the injector in the upper state compared to the parasitic channel from the injector to the lower state. The small energy spacing between the upper and lower radiative states is favoring parasitic injection in the lower state.

4. *Electrical stability.* The current at alignment must be larger than the maximum parasitic current prior to alignment. If this is not the case, the lasing condition would be reached at an electrical instable working point, in the negative differential resistance region, and lasing is not possible. This problem may arise especially for designs with a very efficient extraction from the lower state.

4.2.2 Optical losses

The design of optical confinement structures was a key development in the extension of QCLs from the mid-infrared regime to longer wavelengths. This was due, in part, to the difficulty in scaling the dimensions of conventional dielectric waveguides up with increasing wavelength as well as the increase in free-carrier loss ($\alpha_{fc} \propto \lambda^2$) that occurs in the semiconductor cladding and active regions. Low loss waveguides for the terahertz have been demonstrated, however free-carrier absorption remains a severe issue for low frequency terahertz QCLs.

The Drude theory for the oscillation of an electron driven by an electric field predicts an attenuation proportional to the square of the wavelength for free carrier absorption [108]

$$\alpha_{fc} = \frac{Ne^2\lambda_0^2}{m^* 4\pi^2 n_{op} c^3 \varepsilon_0 \tau} \tag{4.2.1}$$

where N is the electron concentration and n_{op} the refractive index, τ is the relaxation time, λ_0 the free space wavelength, and e the charge of the electron. Fig. 4.2 shows the computed free carrier absorption (assuming a unity confinement factor of the

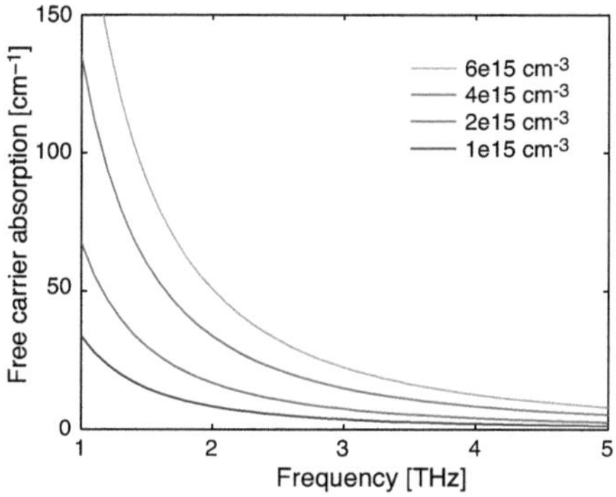

Figure 4.2: Computed free carrier absorption for different doping densities using the Drude model where the relaxation time is $\tau = 0.33$ ps.

optical mode with the active region) as a function of the doping density. A doping density of 4e15 cm^{-3} can be considered as a typical doping density for terahertz QCLs. Clearly for frequencies below 2 THz, the free-carrier absorption losses are higher than the expected maximal available gain in the active region.

The validity of using the Drude model to describe the absorption losses in the active region depends on the active region itself. The majority of electrons reside in the injector region and represent the main contribution to the absorption. If the injector region consists of several subbands that form a miniband, the Drude model of free carrier absorption can be considered as a good approximation. However, if the injector region consists of an isolated subband, the Drude model description is no longer adequate and intersubband absorption should be computed instead. Absorption of the electrons in the injector cannot be avoided, but bandstructure engineering allows to transfer the oscillator strength of the injector intersubband absorption on higher frequencies above the laser frequency. Fig. 4.3(a,b) illustrates the intersubband absorption from an injector, which is part of a miniband. As

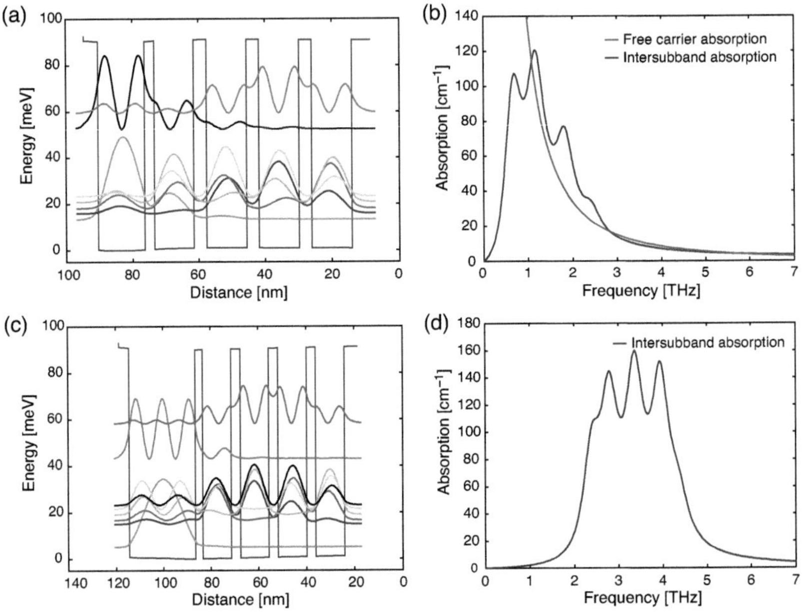

Figure 4.3: (a): An example bandstructure where the injector is the lowest state of a miniband. (b): The computed intersubband absorption assuming all electrons are in the lowest state. The free carrier absorption is a good approximation to the intersubband absorption at low frequencies. (c): A bandstructure where the injector state is separated by an energy gap from the miniband. (d): The computed intersubband absorption assuming again that all electrons are in the lowest state. Both structures have an average doping density of $4.0 \cdot 10^{15}$ cm^{-3}. For the computation of the intersubband absorption a linewidth of 2 meV is assumed and correspondingly a relaxation time of $\tau = 0.33$ ps in the Drude model. The bandstructure in (a), starting from the right is: $3 \times (12.0/\mathbf{3.8})/12.0/\mathbf{3.0}/14.0$. The bandstructure in (c) is identical up to the last well which is 28.0 nm wide. The figures in bold face represent the $Al_{0.1}Ga_{0.9}As$ barriers.

expected, a band with large intersubband absorption up to the energy corresponding to the miniband width is computed. Interestingly, the computed free carrier absorption for the same doping density gives a rather good approximation to the computed intersubband absorption. The key to overcome the large intersubband absorption at low frequency is to separate the injector by an energy offset from the miniband as shown in Fig. 4.3(c,d). The absorption band is shifted by the gap energy to higher frequencies and allows to get low absorption at low frequencies. An active region design must not only address the electron transport and population inversion, but also the design of the intersubband absorption. The latter is especially important for low frequency active regions, where the photon energy is close to the energy spacing of "auxiliary" levels in the injector and relaxation region.

4.2.3 Low frequency bandstructure

In previous terahertz designs, that aimed for operation below 2 THz, the injection efficiency represents a severe issue for the LO-phonon depopulation design, whereas backfilling and intersubband absorption affects the bound-to-continuum approach. The here proposed design for low frequency operation is based on the advantages of a bound-to-continuum scheme, but overcomes the problem of intersubband absorption and back-filling by separating energetically the injector from the miniband. A high injection efficiency is achieved by the split injector.

The bandstructure of the so called "bound-to-continuum with split injector" design for 1.7 THz is shown in Fig. 4.4 [101]. The energy gap splits the injector states from the miniband, allowing a well controlled injection process and reducing the possibility of backfilling. The two injector states are off-resonance prior to the

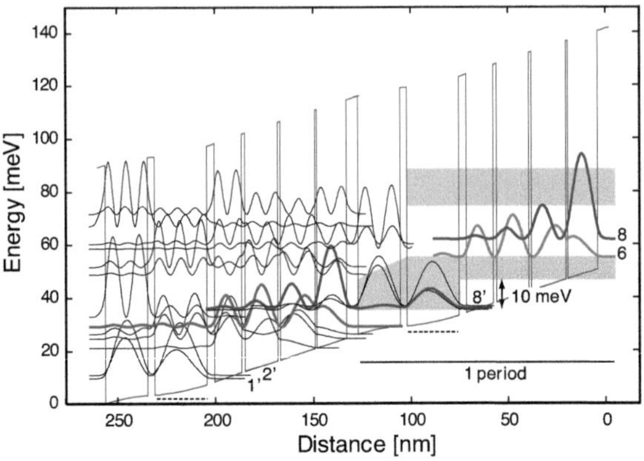

Figure 4.4: Bandstructure for a 1.7 THz design. Self-consistent solution (T = 50 K) of coupled Schroedinger and Poisson's equations for an applied electric field of 2.1 kV/cm. The nominal layer sequence, starting from the injection barrier is (in nm): **5.9**/15.3/**1.0**/17.7/**1.3**/16.6/**1.7**/13.9/**3.9**/26.8/**3.5**/21.4. The figures in bold face represent the $Al_{0.1}Ga_{0.9}As$ barriers and the underlined layer is doped with Si, $1.0 \cdot 10^{16}$ cm^{-3} in density. The doping level yields a sheet carrier density of $2.7 \cdot 10^{10}$ cm^{-2} and an average doping density of $2.1 \cdot 10^{15}$ cm^{-3}.

alignment with the upper state, which reduces efficiently the parasitic current from the injector in the miniband. At injection resonance with the upper state, the injector states themselves are also in resonance, leading to a high injection efficiency in the upper state.

As discussed in the previous section, the intersubband absorption at the laser photon energy is minimized by means of the energy gap. Figure 4.5 shows the calculated intersubband absorption at different temperatures. The model assumes a thermal equilibrium distribution of the electrons starting from the injector. That equilibrium distribution represents a good approximation of the non-equilibrium electron distribution for the computation of the intersubband absorption, since in both cases most of the electrons reside in the injector. Two highly absorbing regions above and below the laser frequency can be identified. They are separated by an absorption

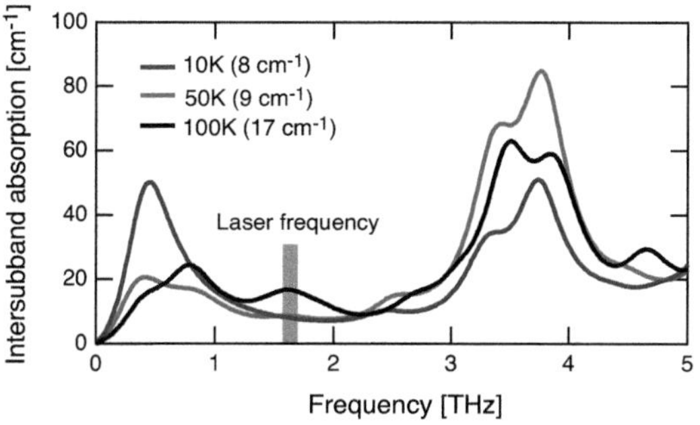

Figure 4.5: Calculated intersubband absorption spectrum of the bandstructure in Fig. 4.4 assuming a thermal equilibrium distribution of the carriers. A Lorentzian broadening of the states with $2\gamma = 2$ meV for the full width at half maximum is assumed. The absorption at the laser frequency is given in brackets.

gap that originates from the energy gap between the injector and the miniband. The absorption above the laser frequency is due to absorption between injector and the miniband, whereas the absorption below the laser frequency is mainly due to absorption between the two injector states. The significant increase of the absorption at the photon energy at 100 K is due to thermally backfilled electrons in the active region.

The computed photon energy between states 8 and 6 is about 7 meV, corresponding to 1.7 THz, the miniband width is 8 meV and the energy gap 10 meV. The dipole matrix element $z_{86} = 10.1$ nm gives an oscillator strength of $f_{86} = 19$. The energy spacing between the two injector states is 1.6 meV and the energy splitting at resonance between injector and upper state is 0.6 meV.

4.2.4 Lifetimes

The advantage of the bound-to-continuum transition is the large phase space for electron scattering out of the upper and lower state of the optical transition, ensuring that the condition for population inversion $\tau_{86} > \tau_6$ is fulfilled [89, 109]. Scattering lifetimes are computed for the bandstructure shown in Fig. 4.4 at alignment bias.

Interface roughness scattering is computed according to Eq. 2.4.41. For the roughness following values are assumed: $\Delta = 0.3$ nm, $\Lambda = 5.0$ nm. Acoustical phonon scattering is computed in the elastic limit, using by Eq. 2.4.39 with $D = 13.5$ eV, $c_l = 1.44 \times 10^{11}$ Nm^{-2} and for a lattice temperature of 10 K. For the alloy disorder scattering Eq. 2.4.43 is used. Ionized impurity scattering is computed according to Eq. 2.4.44 with an isotropic screening model where the screening wavevector is computed in the Debeye approximation, Eq. 2.4.64 with an electron temperature of 70 K ($q_s = 2.15 \times 10^7$ m^{-1}). All scattering lifetimes are computed for an electron with an initial in-plane kinetic energy of 6 meV which corresponds to a temperature of $\sim$ 70 K. Interface roughness scattering lifetimes show a weak in-plane energy dependence, while acoustical phonon and alloy disorder scattering lifetimes don't depend on the in-plane kinetic energy. Table 4.1 summarizes the computed lifetimes.

A direct consequence of the large scattering phase space is that all computed scattering mechanisms are in favor of a population inversion since $\tau_{86} > \tau_6$. As expected, the alloy disorder scattering is very inefficient. Interface roughness scattering is more efficient than acoustical phonon scattering, at a lattice temperature of 10

	ifr	ac	ad	ion	total
τ_8	22	168	251	94	14
τ_6	35	120	395	5	4
τ_{86}	53	364	709	125	32

Table 4.1: Computed scattering lifetimes for interface roughness, acoustical phonon, alloy disorder, and ionized impurity scattering. The lifetimes are in ps units. The electron temperature is 70 K, the lattice temperature 10 K and the scattering rates are computed for an in-plane energy of 6 meV.

K. But acoustical phonon scattering is equally efficient than interface roughness scattering at about 60 K, due to the linear dependence on the temperature. The strength of ionized impurity scattering decreases exponentially with the distance to the impurities. Injector and miniband states that extend into the doping region of the injector are the most affected by ionized impurity scattering.

It is important to note that the above computed lifetimes are the out-scattering lifetimes. The assumed in-plane kinetic energy of the electrons (6 meV) is in the order of the miniband width ($\sim$ 8 meV), therefore back-scattering into the lower state is a serious issue and must be considered for for the computation of the populations.

Scattering by emission of an LO-phonon is not allowed for electrons with an in-plane kinetic energy of 6 meV due to energy conservation. The total energy drop per active region is about 27 meV, therefore the possible intersubband transitions are spaced by less than the LO-phonon energy (36 meV). However hot electrons, with a large in-plane kinetic energy scatter efficiently by emission of an LO-phonon. LO-phonon emission is the most efficient energy relaxation mechanism for the electrons. Due to a simple energy balance argument a probability of 75% per period is estimated for an electron to emit an LO-phonon. Electron-electron scattering is not computed since it is by far the most challenging scattering mechanism to implement. For the doping densities considered in the present structure electron-electron scattering is not likely to be the dominant intersubband scattering mechanism but should be considered for a complete analysis. The authors in Ref. [110] have investigated theoretically and experimentally the electron-electron scattering rates in a quantum well structure in strong magnetic fields. The quantum well structure has a sheet doping of 3.4×10^{10} cm^{-2}, that is comparable to the sheet doping of a terahertz QCL. The extrapolated electron-electron scattering time to zero magnetic field is $\tau_{21}^{ee} \sim 135$ ps.

4.2.5 Terahertz waveguides

For frequencies below 2 THz the corresponding wavelength is longer than 150 μm in the vacuum. Traditional dielectric waveguides are unpractical because of the large cladding thickness required. Solutions rely on the use of metallic layers supporting surface plasmon modes. Two principal schemes have been developed to provide efficient optical resonators for terahertz QCL's: single plasmon and double metal waveguides.

Single plasmon waveguide

This waveguide represented the key ingredient in the demonstration of the first THz QCL [32, 111]. The confinement of the electromagnetic wave is based on a metallic reflection at the top metallization and the quasimetallic confinement provided by a thin, heavily doped buried contact placed below the heterostructure. A two dimensional calculation of the mode intensity is reported in Fig. 4.6(a). The heterostructure has to be grown on a semi-insulating substrate. In this way

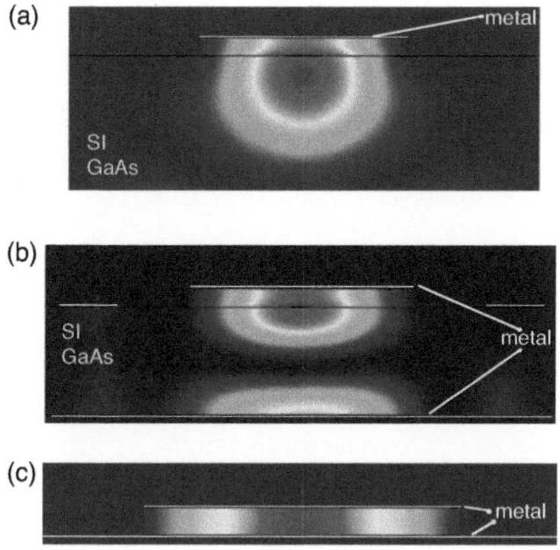

Figure 4.6: Computed mode intensity for (a): a single plasmon waveguide at 3 THz, (b): a hybrid single plasmon waveguide at 1.9 THz with a 600 nm thick buried layer and a thinned down substrate, (c): a double-metal waveguide at 1 THz. For all 3 waveguides, the width of the top metallization is 200 μm.

the large part of the mode which overlaps with the substrate brings a very small contribution to the waveguide losses. The heavily doped (typically $0.8 - 2 \times 10^{18}$ cm^{-3}) buried layer has the double function of optically confine the mode and act as electrical contact. Because of the large and negative dielectric constant of the buried contact, the overlap factor between the field and this heavily doped region is very small. The main advantage of this waveguide resides in its quite high figure of merit (overlap divided by losses) in the $2-4$ THz region in combination with a facet reflectivity still close to the Fresnel one calculated via index mismatch [112] which eases the coupling to free space modes and yields a good far-field and high powers. This waveguide loses its efficiency as the photon energy is lowered: the longer wavelength has to accommodate the same epilayer thickness and the reduction of the dielectric constant of the active region pushes the mode towards the substrate reducing the overlap. Another general disadvantage of this waveguide is the non-optimal and non uniform ridge pumping due to the injection via the buried layer which shows significant in-plane resistance [90]. One possibility to still keep the single plasmon configuration at long wavelength is to rely on a mode which is bounded to the two metallic layers which constitute respectively the top contact and the back metallization of the sample. If the semi-insulating substrate is properly thinned down to a thickness of about a 100 μm (the optimal thickness is wavelength dependent), the mode with the highest figure of merit is the one depicted in Fig. 4.6(b). In this way the active region still experiences a good overlap with the optical mode even if one big part of it is "pulled" towards the lower metallization layer. This waveguide configuration has been successfully employed in the 1.9 THz laser of Ref. [104] and in the device of Ref. [91] at 2 THz. In fact this configuration represents an "hybrid": the two metals guiding the radiation as in a double-metal waveguide described below.

Double metal waveguide

The double metal waveguide is directly derived from the microstrip-line resonator widely employed in the microwave range. In this waveguide the active region is sandwiched between two metallic layers, typically Au, yielding an almost unity overlap factor. It has been first demonstrated in QCL at Mid-IR frequencies [113] and then widely employed in combination of resonant-phonon designs in the devices that have shown the highest temperature performance to-date [62]. Recently, Cu has been proficiently used to enhance the figure of merit of this waveguide setting the new limit in pulsed high temperature operation [20]. The main disadvantage of this waveguide is represented by the patterned far-field [112, 114] and the low

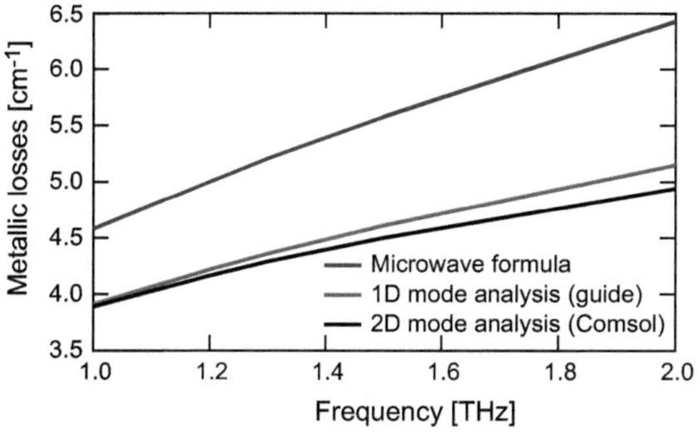

Figure 4.7: Computed waveguide losses of a 200 μm wide and 14 μm thick, empty waveguide. The losses obtained by a 2D and 1D mode analysis are compared to the conduction losses of a microstrip-line using simple analytical microwave formula.

power out-coupling due to the large impedance mismatch at the laser facet. The two-dimensional simulation of the mode intensity for a double metal resonator at 1 THz is plotted in Fig. 4.6(c). The confinement of the optical mode within the active region is close to unity.

Waveguide losses

Double metal waveguides are in terms of the figure of merit the best waveguides for low frequency terahertz QCLs, and employed for all the low frequency lasers in this work. For this reason, only double metal waveguide losses are discussed in this section. The waveguide losses result from the finite conductivity of the metallic layers.

The waveguide losses are computed based on a 2D or 1D mode analysis. The commercial simulation software Comsol is used for the 2D mode analysis and a custom code for the 1D mode analysis. The active region is considered as a lossless dielectric with $n = 3.6$. The gold layers are modeled using measured values for the complex refractive index [115, 116]. The imaginary part of the refractive index accounts for the losses due to the finite conductivity of the metals. The metallic losses decrease at lower frequency due to the higher conductivity of the metals (Fig. 4.7). Mode analysis with a 1D or 2D solver leads to almost identical results for the metallic losses. A simple analytical expression for the conduction losses in microstrip-lines [117], computed for the same waveguide geometry, predicts 20-25% higher losses with a $\sqrt{\omega}$ frequency dependence. For more details on the microstrip-line model, see appendix B.1.

Highly doped contact layers below the metal layers are required to minimize the Schottky barriers for current injection and electrical pumping of the active region. The additional losses due to free carrier absorption could be less than 0.3 cm^{-1} in the 1 to 2 THz range with 100 nm thick contacts, doped to 4×10^{18} cm^{-3}. Note that the lasers discussed in the next sections have thicker contacts, therefore the losses in the highly doped contact layers are slightly larger. Double metal mirror losses are very low due to the large impedance mismatch. They are typically in the range of 1 to 2 cm^{-1} for the considered waveguide geometries and wavelengths.

The total optical losses below 2 THz in an active region embedded in a double-metal waveguide is obtained by the sum of the waveguide losses including free carrier absorption in the contacts (typically $\sim 5 - 10$ cm^{-1}) and the intersubband absorption. The intersubband absorption in the active region of the above presented bound-to-continuum with split injector design are lower than 10 cm^{-1} for a doping density of 2×10^{-15} cm^{-1}. The total optical losses are in the order of 20 cm^{-1} and comparable to those in the $2 - 5$ THz range [112].

4.3 QCL emitting from 1.6 to 1.8 THz

An active region consisting of 120 repetitions of the bandstructure shown in Fig. 4.4 has been grown. The wafer is labeled V303. Laser samples are processed as described in Sec. 3.1.2 and measured in the setup described in Sec. 3.2. The electrical and optical properties of different lasers from the same process are reproduced within typically 10 %, when care is taken that the facets are nicely cleaved, important for getting high output powers of double metal waveguide based lasers.

Laser action is observed between 1.6 and 1.8 THz. The power of a typical sample with Ti/Au contacts measured in pulsed and cw mode is shown in Fig. 4.8(a,b). The highest operation temperature in pulsed mode is 95 K, and for this sample 60 K in cw mode. Another sample of the same process is lasing up to 75 K in cw mode. The maximal cw output power at 10 K is 0.36 mW whereas the peak power in pulsed mode is 0.48 mW. The threshold current density in cw mode at 10 K is 116 A/cm^2.

The inset of Fig. 4.8(b) shows the current-voltage characteristic and the differential resistance. The applied voltage at threshold is 7.6 V and is more than twice the voltage that corresponds to the electric field at injection resonance (2.9 V). The Schottky barriers of the Ti/Au contacts are responsible for the large threshold voltage. Using alloyed contacts a threshold voltage of 3.8 V is measured. The cw performance of a sample with alloyed contacts (Ti/Ge/Au) is reported in Fig. 4.8(c). The lower power dissipation results in a higher cw operation temperature, 80 K for this sample. However the alloyed contacts used in this process are not optimized and introduce additional optical losses which explain the rather poor

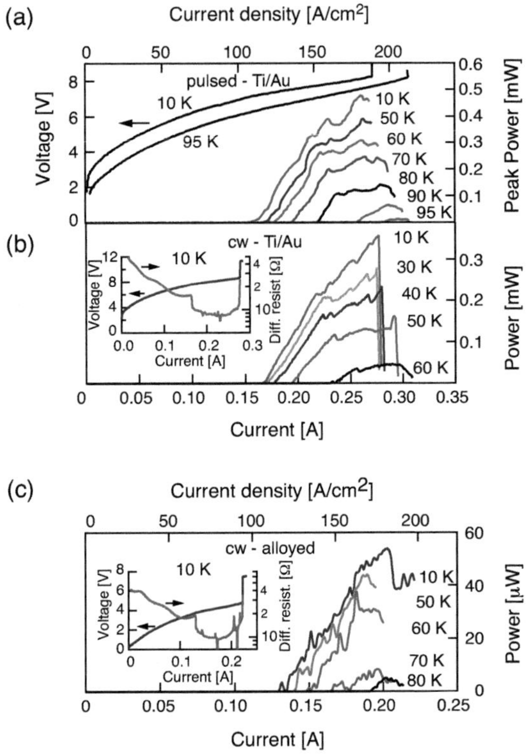

Figure 4.8: (a): Pulsed mode and (b): cw measurements of a device with Ti/Au contacts. The sample is 1.07 mm long and 135 μm wide. The inset shows the voltage-current characteristic and the differential resistance at 10 K. For power measurements in cw mode a calibrated thermopile is used, and for pulsed mode measurement a pyro-electric detector. (c): cw measurement of a device with alloyed contacts. The sample is 1.32 mm long and 85 μm wide. The inset shows the voltage-current characteristic and the differential resistance at 10 K.

power performance.

The emission spectra of all devices show a strong Stark-shift [40] as a function of the applied voltage. Fig. 4.9(a) shows the spectra taken at different current densities of a typical device at 10 K. The center of mass of the emission spectrum tunes from roughly 1.6 THz to 1.8 THz (6.6 to 7.4 meV) with the bias. Lasing occurs on Fabry-Perot modes of the cavity. Due to the very low emission frequency the tuning of the gain curve by the Stark-shift is more than 10 % of the center frequency. Laser emissions on modes in a frequency range of typically 0.1 THz is observed, in agreement with the gain linewidth of about 0.4 THz. The latter is assessed through the measurement of the spontaneous emission linewidth in Fig. 4.9(b). The wider emission spectrum at 160 A/cm^2 could be due to pulse chirp integrated by the low-pass bolometric detection.

Fig. 4.9(c) shows the spectral behavior of a typical sample as a function of the temperature (pulsed mode operation). With increasing temperature the threshold current density increases and hence the tunability decreases. At 50 K the center of mass of the emission energy is still tunable between 1.65 and 1.8 THz. At 95 K the emission spectra is centered at 1.8 THz.

The effective lifetime $\tau_{eff} = \tau_3 \left(1 - \tau_2/\tau_{32}\right)$ is deduced from the equation for the threshold current density [Eq. 2.2.17] by calculating the waveguide losses and the gain cross section

$$J_{th} = \frac{q}{\tau_{eff}} \frac{\alpha}{g_c}$$

where α are the total losses, J_{th} the threshold current density and g_c is the gain

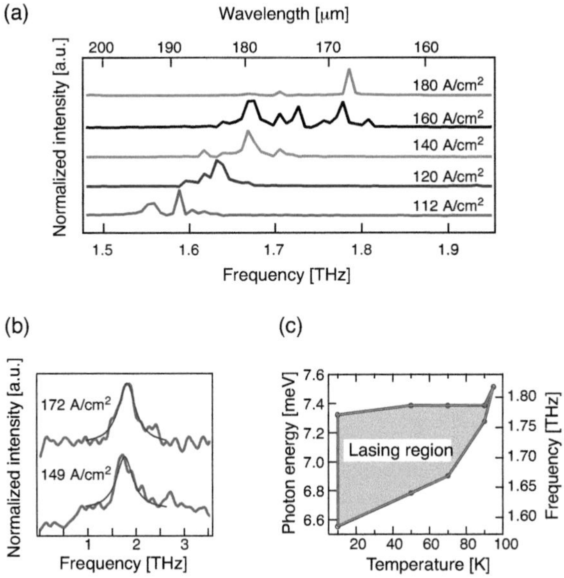

Figure 4.9: (a): Laser spectra at different current densities in pulsed mode for a typical device measured at 10 K. For the measurement a homebuild vacuum FTIR is used. (b): Spontaneous emission spectra from a long-side cleaved single plasmon waveguide ridge. The Lorentzian fit gives a full width at half maximum of 1.5 meV at 149 A/cm^2 respectively 1.6 meV at 172 A/cm^2. (c): Overview of the center of mass of the laser emission spectrum as a function of the temperature in pulsed mode. The colored area shows the accessible lasing frequencies.

cross section given by Eq. 2.2.11

$$g_c = \frac{4\pi\, \Gamma\, q^2\, z_{86}^2}{\varepsilon_0\, n_{eff}\, \lambda\, 2\gamma_{86}\, L_p}$$

where the confinement factor is $\Gamma \sim 1$. Waveguide losses of $\alpha = 19$ cm^{-1} are obtained by summing intersubband absorption in the active region and empty waveguide losses due to the highly doped contact layers and gold contacts. The intersubband absorption in the active region is calculated at 50 K ($\alpha_{ISB} = 9$ cm^{-1}). The empty waveguide losses are $\alpha_e = 10$ cm^{-1}. For the calculation of the gain cross section $2\gamma_{86} = 1.6$ meV is used for the full width at half maximum of the spontaneous emission [Fig. 4.9(b)]. For the effective lifetime $\tau_{eff} = 1.6$ ps is obtained. The measurement of the ratio of the differential resistance above and below threshold ($R_{d,S>0}/R_{d,S=0} = 0.69$) allows to deduce a lower state lifetime of $\tau_2 = 3.5$ ps according to Eq. 2.2.20

$$R_{d,S>0}/R_{d,S=0} = \frac{\tau_2}{\tau_{eff} + \tau_2}$$

The formula for the effective lifetime and the requirement $\tau_{32} \geq \tau_3$ yields an upper bound for the upper state lifetime of 5.1 ps. The deduced lower state lifetime is quite large compared to 1.1 ps of the bound-to-continuum design of reference [90]. The lower aluminium content in the barriers (10% instead of 15%) of the here described structure results in lower interface roughness scattering which could lead to longer lifetimes. The measured slope efficiency extracted from Fig. 4.8(a) is 4.3 mW/A. From the estimated mirror losses [112] of 1 cm^{-1}, a slope efficiency of 6.5 mW/A is calculated using Eq. 2.2.18

$$\frac{dP}{dI} = N_p \frac{h\nu}{q} \frac{\alpha_m}{\alpha_{tot}} \frac{\tau_{eff}}{\tau_{eff} + \tau_2}$$

The collection efficiency of the detector is not taken into account.

The large dynamic range reflects the efficient reduction of the current flow prior to alignment by the two injector states. The dynamic range $(J_{ndr} - J_{th})/J_{ndr}$ of the device shown in Fig. 4.8(a) is 43% and is larger than for other low frequency designs at $\sim$ 2 THz. A dynamic range of 28% is measured for the bound-to-continuum design [91] and 21% for the resonant phonon depopulation design [106].

4.4 QCLs emitting from 1.2 to 1.6 THz

4.4.1 Bandstructure scaling

The bandstructure designed for 1.7 THz and shown in Fig. 4.4 has been scaled to higher frequencies (such as 1.9 THz). More interesting, it has been scaled to lower frequencies and enabled the demonstration of laser operation at frequencies as low as 1.2 THz. The energy spacing between the states is scaled by keeping the same shape of the wavefunctions. Special care is taken for quantum states which are lying close in energy and where the electron transport is dominated by resonant tunneling, which is the case for the two injector states and the upper state at alignment bias. In this case the energy splitting between the individual states is kept identical.

At low photon energy, the doped injector well cannot be narrowed sufficiently by

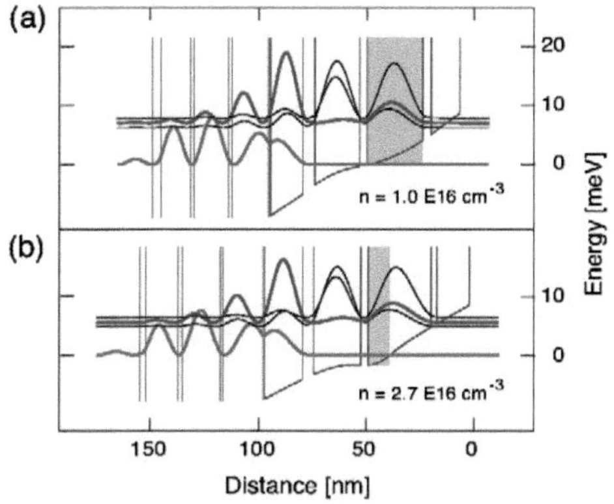

Figure 4.10: Comparison of the injector region for the two different doping schemes. (a): A uniformly doped well is used for designs $\geq$ 1.7 THz and a fractionally doped well is used for designs $\leq$ 1.5 THz. The volume doping density is adjusted to provide the same nominal sheet density.

keeping the coupling scheme of its excited state with the miniband and bringing the two injector states in resonance at the alignment bias. Therefore, the doping scheme is modified for the bandstructure designs below 1.6 THz with the benefit that the average potential drop between the two injector wells is increased, enabling a proper alignment of the two injector states at alignment bias, as shown in Fig. 4.10.

Drawbacks of this doping scheme are that the dopants are closer to the active region, which affects the broadening of the injector state and the upper state lifetime due to ionized impurity scattering [30]. Furthermore the alignment of the injector states and the upper state is more sensitive to variations of the nominal doping during growth and depends on the electron distribution in the quantum states.

4.4.2 Laser characterization

The bandstructure of a laser designed for 1.5 THz is shown in Fig. 4.11(a). The grown sample is labeled N892. The second sample, N908, has an almost identical bandstructure and is designed for an emission of 1.3 THz.

The total losses of 25.2 cm^{-1} for the 1.5 THz laser and 24.5 cm^{-1} for the 1.3 THz laser are higher than the losses of the laser described in the previous section, due to a two times higher measured average doping density compared to the nominal doping density in the active region of $\sim 4 \times 10^{-15}$ cm^{-1}. The waveguide losses, including the highly doped contact layers are 10.3 cm^{-1} for the 1.5 THz and 9.5 cm^{-1} for the 1.3 THz laser. The computed intersubband absorption using the measured doping values of the active region are 14.9 cm^{-1}, respectively 15.0 cm^{-1} for the 1.5 THz, respectively 1.3 THz laser.

Both structures, are grown in the same run and under the same growth conditions by molecular beam epitaxy on GaAs substrates. The grown structures are verified by x-ray measurements. The relative variation of the thickness is -0.3% for the 1.5 THz and -0.9% for the 1.3 THz laser. Figure 4.11(b) shows the excellent agreement between measured and simulated x-ray diffraction of the grown structure N892. Samples are processed into double metal waveguides with non-alloyed Ti/Au contacts, then cleaved, indium soldered on copper mounts and measured on the cold finger of a helium flow cryostat.

Samples from the structure N892 are lasing from 1.34 THz ($\lambda = 224$ μm) to 1.58 THz ($\lambda = 190$ μm) and samples from the structure N908 are lasing from 1.2 THz ($\lambda = 250$ μm) to 1.32 THz ($\lambda = 227$ μm). Together these two structures cover

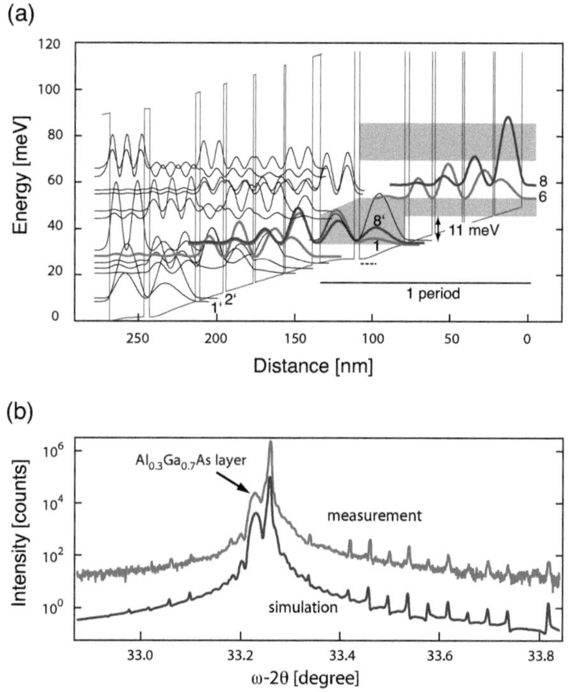

Figure 4.11: (a): Self-consistent solution ($T = 50\ K$) of coupled Schroedinger and Poisson equations for an applied electric field of 1.95 kV/cm for N892. The nominal layer sequence, starting from the injection barrier is (in nm): **5.1**/17.4/**1.0**/18.4/**1.3**/17.4/**1.8**/14.9/**2.9**/19.2/10.0/**3.3**/22.0. $Al_{0.1}Ga_{0.9}As$ barriers are represented by bold face numbers and the underlined layer is doped with Si, $2.7 \cdot 10^{16}$ cm^{-3} in density. The doping level yields a nominal sheet carrier density of $2.7 \cdot 10^{10}$ cm^{-2} and an average doping density of $2 \cdot 10^{15}$ cm^{-3}. (b): Measured and simulated x-ray diffraction of the grown structure (N892).

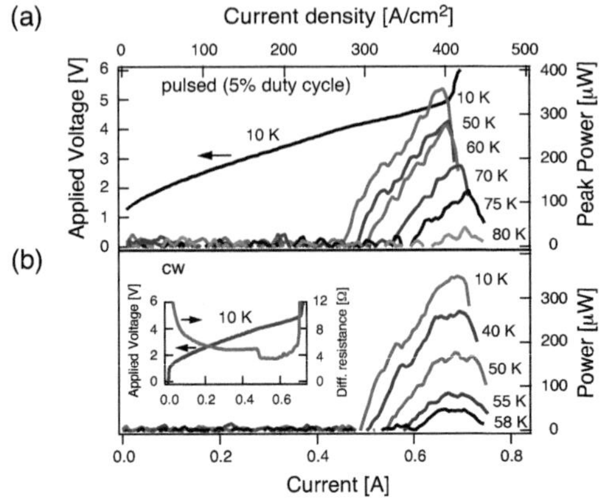

Figure 4.12: (a): Pulsed and (b): cw power-current characteristics of a 1 mm long and 165 μm wide sample of the 1.5 THz laser (N892). The inset shows the voltage-current characteristic and differential resistance at 10 K. For the power measurement an absolute Terahertz power meter is used (Thomas Keating Instruments).

the frequency range from 1.2 THz to 1.6 THz. Up to date, 1.2 THz is the lowest reported operation frequency for a QCL without a strong applied magnetic field [39].

Fig. 4.12 shows the power-current characteristics of a sample of the 1.5 THz laser. At 10 K the cw threshold current density is 286 A/cm^2 and the maximal cw power is 0.36 mW. Operation is observed up to 58 K in the cw mode and up to 80 K in the pulsed mode at a duty cycle of 5%. At a lower duty cycle (0.12%) the sample operates up to 84 K, shown in Fig. 4.14(c). Fig. 4.13 shows the power-current characteristics of a 1.3 THz laser. At 10 K the cw threshold current density is 291A/cm^2 and the maximal cw power is 117 μW. Operation is observed up to 50 K in the cw mode and up to 65 K in the pulsed mode at a duty cycle of 5%. At a duty cycle of 0.12% the sample operates up to 69 K, shown in Fig. 4.14(c). The slope efficiency for the device of the 1.5 THz laser is 1.7 mW/A, and a slope efficiency of

(a)

(b)

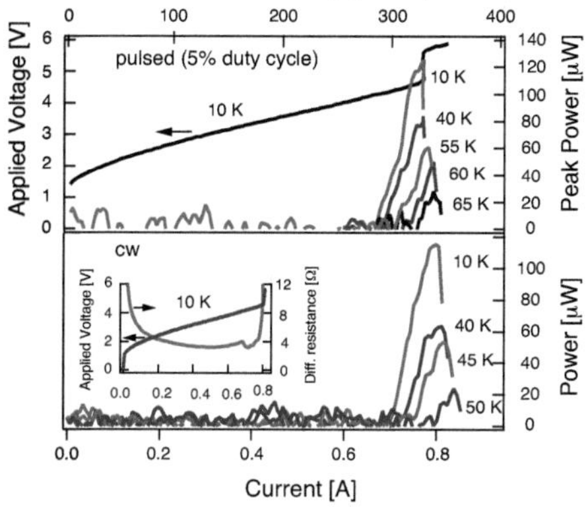

Figure 4.13: (a): Pulsed and (b): cw power-current characteristics of a 1.4 mm long and 165 μm wide sample of the 1.3 THz laser (N908). The inset shows the voltage-current characteristic and differential resistance at 10 K.

1.24 mW/A is obtained for the 1.3 THz laser.

Cw laser spectra for the two devices are reported in Fig. 4.14(a) as a function of the applied bias. The spectra are measured with a custom vacuum Fourier transform infrared spectrometer. Both structures show a strong Stark-shift of the gain curve with the voltage. Lasing occurs on the Fabry-Perot modes of the cavity. The cavity length is chosen to get a wide mode spacing, having for selected voltages only one dominant mode. The strong Stark-shift reveals the diagonal nature of the transition and a relatively low ratio of the upper to lower state lifetime τ_{up}/τ_{dn} [118]. A large ratio of the lifetimes would lead to a photon controlled transport with pinning of the bias voltage above threshold and no significant Stark-shift [119].

The bandstructures of the 1.3 THz laser (N908) and the 1.5 THz laser (N892) are almost identical. The layer sequence of N908 differs only by a couple of monolayers in the first two quantum wells, which are slightly wider (18.2 nm and 18.6 nm

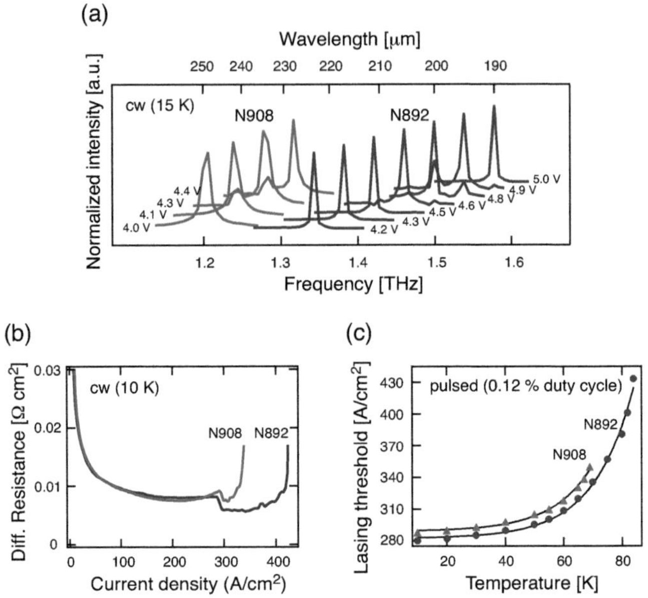

Figure 4.14: (a): Spectral characterization in the cw mode at 15 K of two devices, one from the 1.5 THz laser and the other from the 1.3 THz laser. Both samples are 1 mm long and 165 μm wide. (b): Comparison of the differential resistance in the cw mode at 10 K between the devices of Figure 4.12 and 4.13. (c): Threshold current density versus temperature in pulsed mode of the devices of Fig. 4.12 and 4.13. The solid lines are guides to the eye. The devices are operated at a very low duty cycle (0.12%) and the lasing threshold is detected with a bolometer.

instead of 17.4 nm and 18.4 nm), resulting in a lower lying upper state for N908. Compared to N892, the photon energy of N908 is smaller (4.9 meV instead of 5.7 meV) and the dipole is larger (11.3 nm instead of 10.8 nm), resulting in a only slightly lower oscillator strength (16.6 instead of 17.3).

Apart from the lower lasing frequency no significative difference of the laser and transport properties is expected. Under the assumption of identical intersubband scattering times and linewidth of the spontaneous emission the threshold current density is by Eq. 2.2.17 $J_{th} \propto \alpha_w/g_c \propto \alpha_w L/f_{86}$ where α_w are the waveguide losses,

f_{86} the oscillator strength and L the length of the period. The calculated ratio of the threshold current densities is $J_{th}^{N892}/J_{th}^{N908} = 0.92$, very close to the measured ratio of 0.98. Fig 4.14(b) shows a comparison of the differential resistance of N892 and N908, clearly showing that the transport prior to the lasing threshold and at the threshold is almost identical. The striking difference of the maximal current density before misalignment is in strong contrast to the identical behavior up to the lasing threshold. The lower maximal current density in the 1.3 THz laser is attributed to the smaller anticrossing gap between the injector state and the upper state at injection resonance. The calculated anticrossing gap for the 1.3 THz laser at resonance is $2\hbar\Omega_{N908} = 0.54$ meV compared to $2\hbar\Omega_{N892} = 0.63$ meV for the 1.5 THz laser.

Fig. 4.14(c) shows the threshold current density of both structures as a function of the temperature. The increase of the threshold current density is similar, suggesting that the maximum operating temperature of the 1.3 THz laser is not limited by the photon energy, but by the early misalignment.

4.5 Comparison of lasers from 2.1 - 1.2 THz

A systematic comparison of four different lasers which cover the frequency range from 2.1 to 1.2 THz is presented. The lasers have their center frequency at 1.9 THz (N891), 1.7 THz (N899), 1.5 THz (N892), and 1.3 THz (N908). These four lasers were grown by molecular beam epitaxy during the same run under the same growth conditions, which is an advantage for comparison. Table 4.2 on page 112 gives an overview on the four lasers. The layer N899 is similar in design and performance to the layer V303 which was grown in a different chamber by H. Beere in Cambridge [101]. The experimental data of layer N892, N908 and V303 have been discussed in detail in the two previous sections.

4.5.1 Frequency coverage

The 1.9 THz laser and 1.7 THz are based on the bandstructure with a uniformly doped injector well, scaled to the desired photon energy. The 1.7 THz laser is a bi-stack structure with the designed gain curves of the two stacks shifted by 0.6 meV. The 1.5 THz and 1.3 THz laser are based on the bandstructure with the fractionally doped injector well, discussed in the previous section. Fig. 4.15(a) shows an overview of cw spectra of the four lasers. For each laser different spectra are measured at different voltages. Although the gain curve of the lasers can be widely and continuously tuned by the Stark-shift, the lasing frequency is fixed by the optical modes of the cavity. The gain curve of each laser can be tuned in such a way that these four lasers cover together the frequency range from 2.1 to 1.2 THz. For the chosen dimensions of the cavities, the 1.5 THz and 1.3 THz lasers operate mostly on only one longitudinal cavity mode contrary to the 1.7 THz and 1.9 THz laser that show multimode operation for most of the voltages. Higher lateral modes in the 1.7 THz and 1.9 THz laser have low enough losses to lase due to the shorter wavelength.

4.5.2 Emitted power

Fig. 4.15(b) gives an overview on the measured power as a function of frequency of four lasers. The power is measured at 10 K in the cw mode. The optical cw

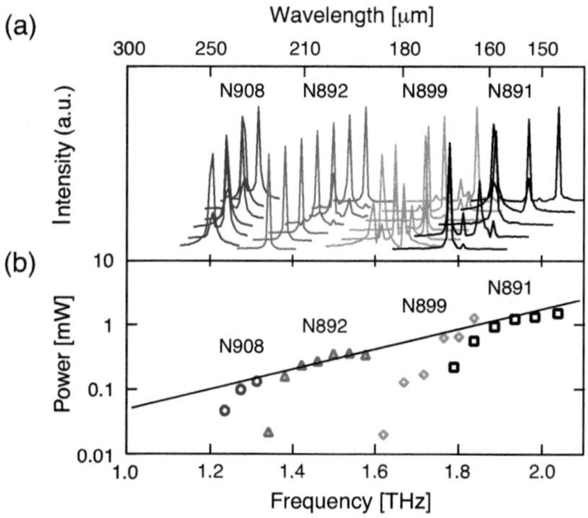

Figure 4.15: (a): Spectral characterization in the cw mode at 10 K of four devices. Spectra are measured for different applied bias. The devices are all 1 mm long, and 165 μm wide, except the one of the 1.7 THz laser which is 110 μm wide. (b): Measured power of four lasers in cw mode at 10 K as a function of the frequency. All lasers are 1 mm long and 165 μm wide. For the power measurement on absolute terahertz power meter is used (Thomas Keating Instruments).

power at 10K is only slightly lower than the peak power in pulsed mode. The power generated from each laser shows a strong frequency dependence due to the Stark-shift. The line indicates the trend of lower achieved power at lower frequency which results from a combination of a smaller slope efficiency, a smaller dynamic range and smaller mirror losses at lower frequencies.

The slope efficiency describes the efficiency of photon generation by pumping electrons through the laser and the out-coupling of the photons. It is given by Eq. 2.2.18

$$\frac{dP}{dI} = N_p \frac{h\nu}{e} \frac{\alpha_m}{\alpha_w + \alpha_m} \frac{\tau_{eff}}{\tau_{eff} + \tau_2}$$

where $\tau_{eff} = \tau_3(1 - \tau_2/\tau_{32})$. In table 4.2 the measured values of the slope efficiency for these lasers are listed. There is a difference by roughly a factor of 4 between the slope efficiencies of the lasers lasing above 1.6 THz and those lasing below 1.6 THz.

The slope efficiency is proportional to the photon frequency and to the number of periods. The 1.5 THz and 1.3 THz lasers have significantly fewer periods, giving a lower slope efficiency due to less gain, but also due to the lower photon energy. Fewer periods results also in a thinner active region, and due to the subwavelength-confinement of the mode to a larger impedance mismatch between waveguide and free space, leading to higher facet reflectivity and by consequence lower mirror losses. In the considered regime of active region thickness and frequency, the slope efficiency has a strong dependence on the number of periods, but it is an easily adjustable parameter. For example the slope efficiency of the 1.5 THz laser is expected to be 1.7 times larger by only growing 25 additional periods, which would result in a total thickness of 15.4 μm of the active region. Independently from the thickness of the active region in the considered range, the impedance mismatch between waveguide and free space remains very high, resulting in the main limiting factor for the slope efficiency. For the whole set of devices the mirror losses are typically 10 times lower than the waveguide losses (Tabel 4.2). By improving the impedance matching, an order of magnitude higher slope efficiencies could potentially be achieved with the same active region.

The internal quantum efficiency $\eta_{int} = \tau_{eff}/(\tau_{eff}+\tau_2)$ is deduced from the measured slope efficiency. The internal quantum efficiency doesn't show a significant decrease with frequency (Table 4.2).

The dynamic range affects the total power which is generated by the laser. It is defined by

$$D_r = \frac{J_{max} - J_{th}}{J_{max}} \tag{4.5.2}$$

The dynamic range of the 1.5 THz and especially for 1.3 THz laser are significantly smaller than those of the 1.7 THz and 1.9 THz laser (Table 4.2), also contributing to the lower total power obtained at lower frequency.

4.5.3 Temperature performance

The threshold current density of the four lasers as a function of the temperature is shown in Fig. 4.16(a). All four lasers show a negligible increase of the threshold current density up to 50 K, followed by a relatively strong increase up to the maximal operation temperature. The highest operation temperature of 100 K is reached by the 1.9 THz laser.

The threshold current density normalized to the maximal current at injection res-

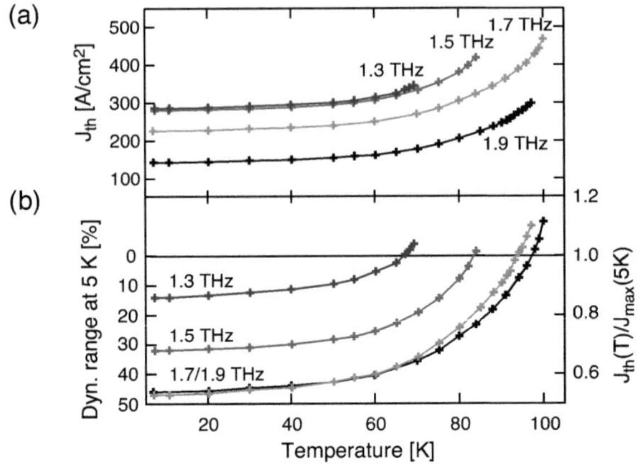

Figure 4.16: Temperature performance of four lasers. (a): Threshold current density as a function of temperature measured in pulsed mode at a very low duty cycle (0.12%). The lasing threshold is detected with a bolometer. (b): The threshold current density normalized by the maximum current density J_{max} at 5 K . The dynamic range at 5 K is displayed on the left axis.

onance J_{max} at 5 K is shown in Fig. 4.16(b). It allows a more easy comparison among the four lasers which have different threshold current densities and dynamic ranges. The dynamic range at 5 K can be read on the left axis. The normalized threshold current density of the different lasers shows a similar increase with temperature. There is an apparent link between the dynamic range at 5 K and the maximal operating temperature. The graph suggests that the lower maximal operating temperature of the lasers at lower frequency is rather related to the smaller dynamic range at 5 K than to the lower photon energy. The maximal current J_{max} increases with the temperature therefore the normalized threshold current density exceeds unity at higher temperatures. At the highest operating temperature of the 1.9 THz and 1.7 THz laser, the maximal current J_{max} increases by about 12%, indicating that in these lasers the upper state lifetime decreases with the temperature

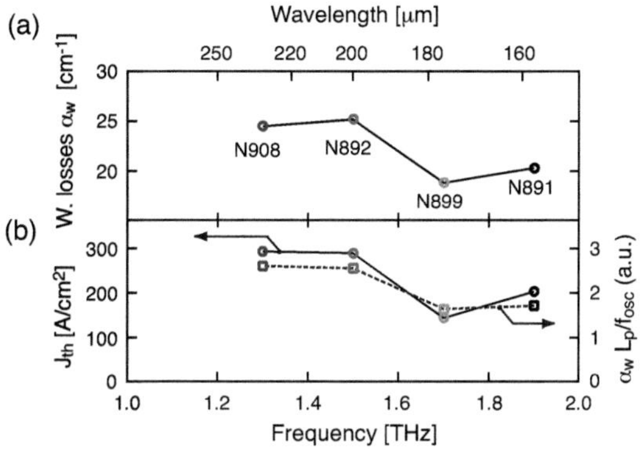

Figure 4.17: (a): Calculated waveguide losses α_w. (b): Comparison between the measured threshold current density (left axis) and the calculated expression $\alpha_w L_p / f_{osc}$ (right axis).

(see Sec. 2.2.5).

The reason for the increase of the threshold current density with temperature is still not well understood and could be due to different mechanisms. Decrease of the upper state lifetime may be a reason, but also the broadening of the states could increase with temperature due to higher in plane scattering which would lead to a larger linewidth, lower gain cross section and a lower injection selectivity. The intersubband absorption and free carrier absorption might as well increase with the temperature. In Sec. 6.3.2 indications for a temperature activated current leakage channel prior to the alignment is found. At higher temperature the onset of injection in the upper state occurs at larger current densities, reducing therefore the dynamic range with temperature.

4.5.4 Waveguide losses

Waveguide losses are calculated for the four lasers according to the description in Section 4.2.5. For the calculation of the intersubband absorption the doping values measured by capacitive measurements [120] and reported in Table 4.2 are used. The measured average doping in the active region is about twice the intentional doping. Fig. 4.17(a) shows the calculated waveguide losses for all four lasers. Due to loss engineering via the bandstructure and the waveguide, the losses do not show a λ^2 increase with the frequency, which would be expected if free carrier losses were dominating in the structure. There is still a slight, but not dramatic, increase of the losses with the wavelength.

The threshold current density depends on the waveguide losses. In a three level model the threshold current density is given by Eq. 2.2.17

$$J_{th} = \frac{e}{\tau_{eff}} \frac{\alpha_{tot}}{g_c}$$

where the total losses α_{tot} are the sum of waveguide and mirror losses and g_c is the gain cross section. Under the assumption of identical intersubband scattering times and linewidth of the spontaneous emission, the threshold current density J_{th} reads, $J_{th} \propto \alpha_w L_p / f_{osc}$, where α_w are the waveguide losses, f_{osc} the oscillator strength, and L_p the length of the period. The mirror losses are one order of magnitude smaller than the waveguide losses and are neglected. The computed values of $\alpha_w L_p / f_{osc}$ are compared in Fig. 4.17(b) to the measured threshold current densities, showing a good qualitative agreement. For the lasers below 1.6 THz a significantly higher threshold current density is predicted than for those above 1.6 THz, in full agree-

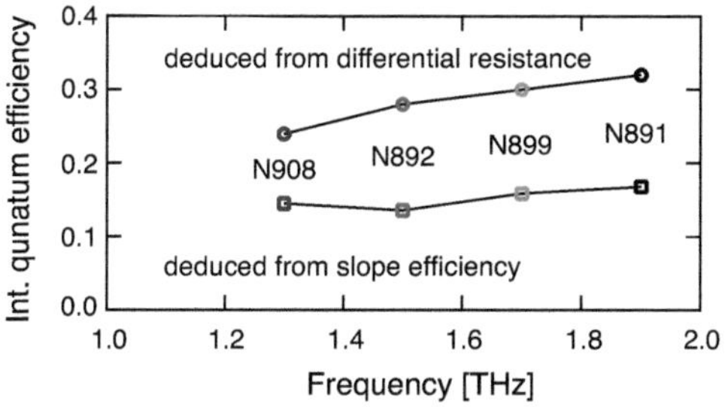

Figure 4.18: Deduced lower limit of the internal quantum efficiency at threshold, by the measurement of the slope efficiency and the change of the differential resistance at threshold.

ment with the measurements. The higher threshold is a combination of higher losses and lower oscillator strength, underlining the importance of optimizing the waveguide losses. However this simple model shows a qualitative discrepancy for the threshold current density of the 1.7 THz laser when compared to the 1.9 THz laser. This difference in the threshold current density of these two lasers might be related to the injection efficiency, as discussed further in Sec. 4.5.6.

4.5.5 Lifetimes

The laser performance depends crucially on the lifetimes of the intersubband levels. Although lifetimes are not directly measurable, the lifetime ratio $\tau_{eff}/(\tau_{eff}+\tau_2)$, which represents the internal quantum efficiency of the laser transition, can be deduced from measurements. The slope efficiency is related to the internal quantum efficiency by Eq. 2.2.18 and its value can then bring informations on the lifetime ratio. Another possibility is offered by the measurement of the differential resistance. The ratio of the differential resistance above and below threshold is related to the internal quantum efficiency by Eq. 2.2.20

$$R_{d,S>0}/R_{d,S=0} = 1 - \frac{\tau_{eff}}{\tau_{eff} + \tau_2}$$

Both methods allow to deduce a lower limit of the internal quantum efficiency, shown in Fig. 4.18. An additional series resistance would lead to an underestimation of the internal quantum efficiency by the technique of the differential resistance, whereas the non-unity collection efficiency of the power measurement setup leads to an underestimation by the technique of the slope efficiency. Taking the data obtained from the differential resistance as a reference, a power collection efficiency of 50% is estimated for the optical setup.

The internal quantum efficiencies obtained by both methods are plotted in Fig. 4.18 and decrease as the laser frequency is lowered; the population inversion degrades. However the decrease is not dramatic and doesn't scale with the energy spacing between upper and lower state which varies from 8 to 5 meV for the considered lasers. Still a sizeable population inversion should be achievable at even lower frequencies if the injection selectivity is guaranteed.

4.5.6 Injection efficiency

The electronic injection efficiency in the upper lasing level is not accessible from a direct measurement. However a careful analysis of the lasers transport properties gives indications on the injection efficiency. Fig. 4.19(a) shows the current-voltage ($I - V$) characteristics of the four lasers. The large discontinuity in the $I - V$ characteristics is due to the region of negative differential resistance after the in-

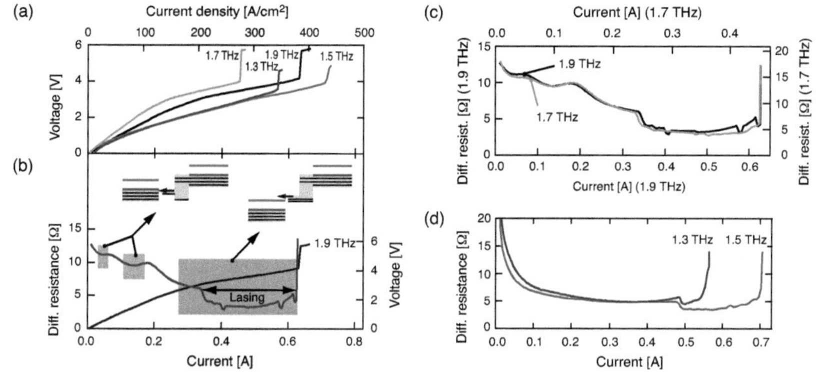

Figure 4.19: Comparison of transport properties of four lasers. The same lasers as in Fig. 4.15(b) are used, the dimension of all lasers are 1 mm x 165 μm. (a): $I-V$ curves of the lasers (a constant voltage drop due to Schottky barrier has been substracted). (b): $I-V$ curve of the 1.9 THz laser and its differential resistance, obtained by a differentiation of the $I-V$ curve. Different transport regimes are identified. At low current, electrons are injected in miniband states, at higher current electrons are injected in the upper state. (c): Comparison of the differential resistance of the 1.9 THz and 1.7 THz lasers and (d): differential resistance of the 1.5 THz and 1.3 THz lasers.

jection resonance. The differential resistance in the low voltage regime is not very steep, especially for the 1.5 THz and 1.3 THz lasers, indicating an strong electron transport through the structure prior to alignment.

The differential resistance is obtained by a differentiation of the $I-V$ curve and is shown for the 1.9 THz laser in Fig. 4.19(b). The differential resistance reveals two humps in the low current regime due to injection resonances from the injector state to miniband states. The alignment of the injector state with the upper state starts after the second hump. In this picture, the injection efficiency in the upper state vanishes in the low current regime and increases only after the second hump. Fig. 4.19(c) shows a comparison of the differential resistance of the 1.9 THz and 1.7 THz lasers. To compare the two lasers, since each one has a different current density range, separate x and y-axis are used. The x-axis of the 1.7 THz laser is scaled by 0.72 and accordingly, the y-axis by 1/0.72, compared to the 1.9 THz laser. The almost perfect overlap of the features in the differential resistance and threshold suggest nearly identical lifetimes and transport regimes in the two structures. Since the threshold current densities do not follow the prediction (see Fig.4.17) it may be argued that the threshold condition in this case is affected by the injection efficiency leading to voltage-controlled device [121]. This may explain the quite different threshold current densities displayed by these two lasers ($J_{th}^{1.9THz}$ = 203 A/cm^2, $J_{th}^{1.7THz}$ = 144 A/cm^2). The scaling factor for the current which is used in Fig. 4.19(c) is mainly due to the doping which is indeed 20% lower in the 1.7 THz than in the 1.9 THz laser.

The differential resistances of the 1.5 THz and 1.3 THz lasers, shown in Fig. 4.19(d), look very similar. The lower maximal current density of the 1.3 THz laser is attributed to the smaller energy splitting $2\hbar\Omega$ between the injector state and the upper state at injection resonance, and possibly also to the 10% lower doping in the 1.3 THz laser, resulting in a smaller dynamic range. The absence of humps together with the less steep $I-V$ curves and lower differential resistance in the low current regime point towards a large non-resonant scattering channel from the injector state to the miniband prior to alignment. The overall injection selectivity of the injector is reduced in the 1.5 THz and 1.3 THz laser compared to the lasers at 1.7 THz and 1.9 THz. A gradual decrease of the injection selectivity with lower frequencies is expected due to lower energy spacings between the electronic states. However the very different transport properties of the 1.5 THz and 1.3 THz lasers suggest that the abrupt decrease of the injection selectivity is rather related to the modified doping profile than to the downscaled energies. The doping in the 1.5 THz and 1.3 THz lasers, is closer to the injector well and active region. Stronger intra-subband ionized impurity scattering in the injector leads to a broader injector state and might be responsible for a lower injection selectivity in the upper state. Also, during the growth of the structures the dopants migrate in the growth direction resulting in a shift and smearing out of the doping profile [122]. The alignment of the bandstructure crucially depends on the self-consistent calculation of the potential with the nominal doping profile. A different effective doping profile may affect the proper alignment of the injector states.

4.6 Application of a 1.5 THz QCL

Heterodyne receivers up to about 1.5 THz [123] or 2 THz [124] have been demonstrated based on solid state multiplier chains as local oscillators. At higher frequencies, only local oscillators based on optically pumped gas lasers [125], limited to selected, fixed frequency operation, or multiplied backward-wave oscillator sources [126], limited to a narrow tuning range, have demonstrated successful operation in heterodyne receivers. The high output power ($\geq$ 100 μW) makes QCLs very attractive as potential local oscillators, in particular for array receiver applications. A 1.5 THz quantum cascade laser (QCL) has been investigated as a local oscillator in the context of the development of terahertz receivers for the Stratospheric Observatory for Infrared Astronomy (SOFIA) [127], through a collaboration between ETH and the Universität zu Köln [128]. Phase locked QCL operation and its use as a local oscillator source in a heterodyne receiver have been demonstrated [129]. The reference signal for the QCL phase lock is generated by a phase locked Gunn

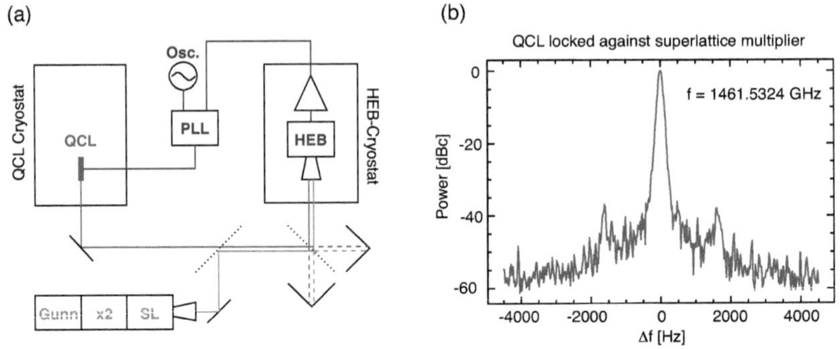

Figure 4.20: Figure reprinted from Ref. [128]. (a): Phase lock of a QCL at 1.46 THz. The QCL is mixed on a hot electron bolometer with a reference signal produced by a Gunn oscillator followed by a Schottky frequency doubler and a superlattice harmonic generator. (b): The line width in the spectrum is limited by the resolution bandwidth (100 Hz) of the spectrum analyzer.

oscillator at 121.7 GHz, followed by a conventional Schottky frequency doubler and a superlattice device, which creates the 12th harmonic of the Gunn signal near 1460 GHz [Fig. 4.20(a)]. This signal is mixed with the QCL on the hot electron bolometer. The beat signal is compared with a quartz oscillator in a digital PLL circuit, which produces the drive voltage for the QCL. Due to its smooth voltage tunability, the laser can be stabilized to a line width $\leq$ 100 Hz [Fig. 4.20(b)]. The receiver noise temperature with the QCL local oscillator is identical to the noise temperature obtained with a frequency multiplier chain local oscillator.

4.7 Conclusions

A bound-to-continuum bandstructure with split injector has been developed for low frequency terahertz QCLs. The main advantage of this bandstructure is the low intersubband absorption at the lasing frequency due to the minigap. The large scattering phase space due to the miniband results in a favorable condition for population inversion for all considered scattering mechanisms. A good injection efficiency at alignment bias is obtained due to the split-injector.

A series of terahertz QCLs has been developed to cover the frequency range from 2.1 to 1.2 THz, and demonstrating the lowest operation frequency of a terahertz QCL without a strong applied magnetic field. The identified limiting mechanism for scaling to lower frequencies is the injection efficiency. The lower energy spacing between the upper and lower state leads to a decrease of the injection efficiency for lower frequencies. In addition, the modified doping profile used for the QCLs below 1.6 THz is suspected to increase the broadening of the injector state resulting in a strong non-selective leakage channel prior to alignment. No indications for a drastic increase of the optical losses, nor a dramatic decrease of the internal quantum efficiency, that reflects the lifetimes, are found.

Frequency, THz	1.9	1.7	1.5	1.3
Layer	N891	N899	N892	N908
Energy splitting $2\hbar\Omega$, meV	0.67	0.67	0.63	0.54
Thickness of AR, μm	14.2	15.9	12.1	12.2
Number of periods	110	2×60	85	85
Period length, nm	122.8	126.8	134.7	135.7
Oscillator strength	18.9	18.9	17.3	16.6
Carrier density, 10^{10} cm^{-2}	2.6	2.2	2.7	2.7
Avg. doping, 10^{15} cm^{-3}	2.1	1.7	2.0	2.0
Facet Reflectivity	0.80	0.80	0.86	0.88
$\alpha_{ISB}^{T=50K}$, cm^{-1}	12.0	10.5	14.9	15.0
α_w^{empty}, cm^{-1}	8.3	8.3	10.3	9.5
W.guide losses α_w^{tot}, cm^{-1}	20.3	18.8	25.2	24.5
Mirror losses, L=1 mm, cm^{-1}	2.2	2.2	1.5	1.3
Slope efficiency, mW/A	7.2	7.1	2.0	1.7
Threshold current, A/cm^2	203	144	289	293
Maximum current, A/cm^2	380	273	426	341
Dynamic range	0.46	0.47	0.32	0.14
Stark-shift of gain, meV	1.0	1.0	1.0	0.6
T_{max}, K	100	97	84	69
Avg. doping, 10^{15} cm^{-3}	4.06	3.45	3.99	3.61
$R_{d,S>0}/R_{d,S=0}$	0.68	0.70	0.72	0.76
Internal quantum efficiency	0.17	0.16	0.14	0.15

Table 4.2: Summary of some characteristics of the four bound-to-continuum lasers with split injector. Calculated values and design parameters are shown above the double line, measured and deduced values below. The internal quantum efficiency is deduced from the measured value of the slope efficiency.

Chapter 5

Magneto-transport of low frequency QCLs

5.1 Magneto-transport

Already before the demonstration of the first QCL, it has been suggested that the application of a magnetic field perpendicular to the quantum well layers could dramatically reduce the non-radiative scattering rate of intersubband transitions. Upon the application of the magnetic field, the in-plane parabolic dispersion relation of the subbands breaks into discrete sets of equidistant Landau levels. Thus, when the magnetic field strength is chosen appropriately, it is possible to forbid the various non-radiative intersubband transition processes due to the new energy conservation requirement [130]. On the other hand, the intersubband scattering rate shows a resonant enhancement if the Landau levels corresponding to different subbands are aligned. The first experimental demonstration of elastic transport resonances in a superlattice that stem from the interplay between electric and magnetic field is presented in Ref. [131]. In these experiments, elastic scattering with ionized background impurities and with layer fluctuations are the main mechanisms assisting resonant transport in the low temperature range. In an analytical theory it was shown that resonant intersubband scattering associated with the alignment of staircases of Landau levels corresponding to different subbands results in magneto-intersubband oscillations of the conductivity of a two dimensional electron gas occupying two subbands. The required large moment transfer for intersubband scattering is enabled by short-range disorder, such as interface roughness or impurities, close to the electron gas [132].

5.1.1 The Hamiltonian

A multi-quantum well system immersed in a perpendicular magnetic field (along the z direction) is shown in Fig. 5.1(a). The Hamiltonian for the electrons writes as [40]

$$\left(\frac{(\mathbf{p}+e\mathbf{A})^2}{2m^*} + V(z) + g^*\mu_B\sigma B\right)\Psi(x,y,z) = E\Psi(x,y,z) \tag{5.1.1}$$

σ is the Pauli operator for the spin, μ_b is the Bohr magneton and g^* the effective Landé factor. The Landau gauge is employed for the vector potential $\mathbf{A} = (-By, 0, 0)$, corresponding to a magnetic field along the z direction $\mathbf{B} = (0, 0, B)$. It turns out that the solution can be written as $\Psi(x,y,z) = \psi(z)\Phi(x,y)$, where $\psi(z)$ the solution of the one-dimensional Schrödinger equation without magnetic field

$$\left(\frac{-\hbar}{2m^*}\frac{d^2}{dz^2} + V(z)\right)\psi(z) = E\psi(z) \tag{5.1.2}$$

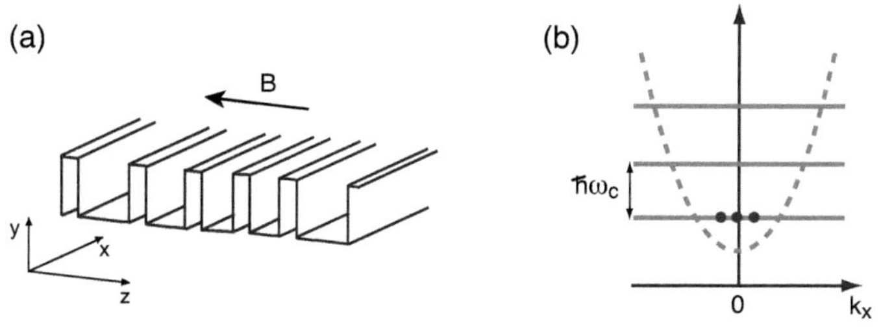

Figure 5.1: (a): Multi-quantum well system in a strong perpendicular magnetic field. (b): The parabolic in plane dispersion relation breaks up into a set of equidistant Landau levels in the magnetic field.

The solution for $\Phi(x, y)$ is expressed with the Hermite polynomials H_n in the y direction

$$\Phi(x,y) = e^{ik_x x}(2^n n! \sqrt{\pi} l_B)^{-1/2} e^{\left(-\frac{(y-y_0)^2}{2l_B^2}\right)} H_n\left(\frac{y-y_0}{l_B}\right) \tag{5.1.3}$$

where $l_B = \sqrt{\hbar/(eB)}$ is the magnetic length that represents the width of the ground state of a Landau level. The Landau levels form a discrete energy spectrum and are given by

$$E_n = E_z + \hbar\omega_c\left(n + \frac{1}{2}\right) \pm \frac{1}{2} g^* \mu_B B \qquad n = 0, 1, 2, ... \tag{5.1.4}$$

where $\omega_c = eB/m^*$ is the cyclotron energy. For GaAs $\hbar\omega_c/B = 1.73$ meV/T. Non-parabolicity effects are neglected in this expression and also the spin degree of freedom is neglected in the following, since in GaAs $g^* = -0.44$ and therefore the spin splitting at 12 T is only about 0.6 meV, well inside the broadening of the

Landau levels. Therefore

$$E_n = E_z + \hbar\omega_c\left(n + \frac{1}{2}\right) \qquad n = 0, 1, 2, ... \tag{5.1.5}$$

Fig. 5.1(b) illustrates the continuous energy spectrum of a parabolic subband that breaks into a set of equidistant Landau levels in the magnetic field.

5.1.2 Transport in the magnetic field

In the magnetic field, the scattering rate between Landau levels of corresponding subbands is quenched. The energy conservation requirement reduces the number of possible scattering processes due to the discrete energy spectrum in the magnetic field. However for particular values of the magnetic field, corresponding to Landau level resonances, the scattering rate is enhanced. At resonance, the cyclotron energy $\hbar\omega_c$ is related to the corresponding subband energy spacing ΔE_{21} by

$$\Delta E_{21} = n\hbar\omega_c \qquad n = 1, 2, 3, ... \tag{5.1.6}$$

Fig. 5.2(a,b) shows the situation where the elastic scattering is quenched due to an off-resonance condition, and the situation where elastic single electron scattering is allowed, corresponding to a Landau level resonance. It is important to note that at resonance coherent tunneling of levels with different Landau indices is forbidden, while tunneling assisted by elastic scattering is allowed [133].

The application of a perpendicular magnetic field to a QCL allows to manipulate the scattering rates of the intersubband transitions. The upper level of the lasing

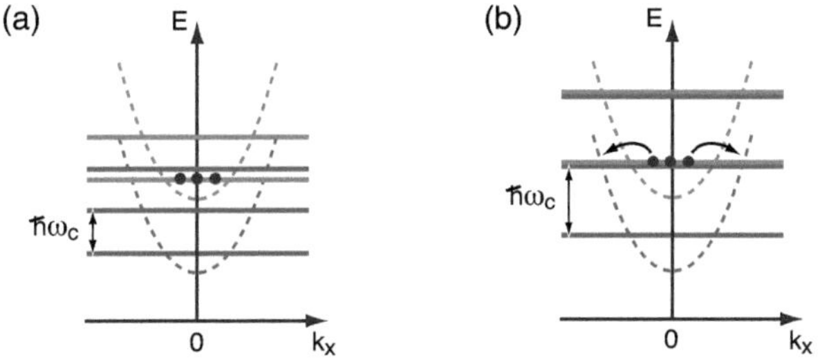

Figure 5.2: (a): The Landau levels are not in resonance therefore elastic single electron scattering between the two quantum states is not allowed. (b): Landau level resonance with $n = 1$, where elastic single electron scattering between Landau levels of the two quantum states is allowed.

transition represents a bottleneck for the electron transport in the active region due to its long lifetime. The lifetime modulation of the upper subband by the magnetic field results therefore in particularely strong modulations of the total current through the QCL active region. The study of the current variations as a function of the magnetic field at fixed bias allows a spectroscopical inspection of the energy levels and contains informations on the scattering mechanisms in the QCL active region. Furthermore the magnetic field can be used to enhance the upper state lifetime through an off-resonance condition.

Magnetic field enhanced emission has been observed in terahertz quantum cascade emitters before the demonstration of the first terahertz QCL [134, 135]. Pronounced oscillations in output power and threshold current density is observed in terahertz QCLs in perpendicular magnetic fields [136]. For particular values of the magnetic field a reduction of the threshold current density and a simultaneous enhancement of the laser emission intensity are observed in Ref. [76, 137] and many others. QCLs with threshold current densities as low as 1 A/cm^2 have been demonstrated in high magnetic fields [102, 104]. In those QCLs, the active region is based on a wide quantum well and population inversion is achieved between the second and first excited state. The upper state lifetime is increased by an off-resonance condition, while the lower state lifetime is simultaneously decreased by a Landau resonance condition. In Ref. [138] laser operation up to 225 K at 3 THz of a resonant phonon depopulation design is observed through the suppression of non-radiative scattering in strong magnetic fields.

Studies of the laser properties and transport in the magnetic field allows to highlight the different scattering mechanisms in QCLs. The emitted power and magnetoresistance as a function of the magnetic field shows a modulation of the electron lifetime due to inelastic and elastic scattering mechanisms [71]. The role of interface roughness scattering in bound-to-continuum structures is pointed out in Ref. [76, 139] and magneto-transport measurements provide evidence that elastic scattering limits the lifetime of a four well terahertz QCL at low temperatures [75].

In the present work the magnetic field is used as an analytical tool for the study of QCL active regions. Magneto-transport studies of bound-to-continuum based lasers allow to deduce the relative injection efficiency in the upper state of the lasing transition. On non-lasing bound-to-continuum structures, magnetic field studies of the Landau level resonances allow to verify the transition energy and to reach laser operation at high magnetic fields through the enhancement of the upper state lifetime.

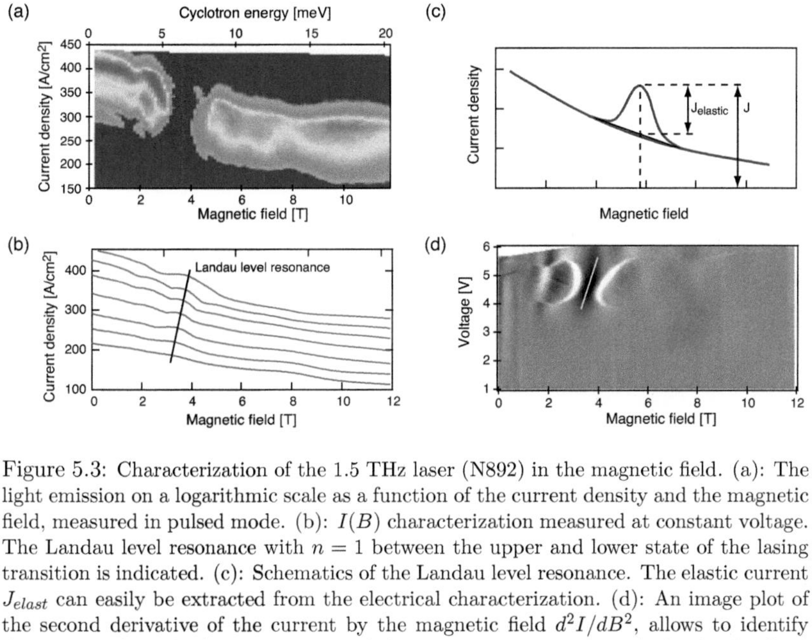

Figure 5.3: Characterization of the 1.5 THz laser (N892) in the magnetic field. (a): The light emission on a logarithmic scale as a function of the current density and the magnetic field, measured in pulsed mode. (b): $I(B)$ characterization measured at constant voltage. The Landau level resonance with $n = 1$ between the upper and lower state of the lasing transition is indicated. (c): Schematics of the Landau level resonance. The elastic current J_{elast} can easily be extracted from the electrical characterization. (d): An image plot of the second derivative of the current by the magnetic field d^2I/dB^2, allows to identify the regime where the laser operates, and the transition energy through the Landau level resonance condition.

5.1.3 Elastic current

The 1.5 THz laser (N892) has been characterized in the magnetic field. Fig. 5.3(a) shows the measured power as a function of the threshold current density and the magnetic field. The decrease in the threshold current density with increasing magnetic fields is similar to what has been measured on other bound to continuum structures in Ref. [76]. The suppression of lasing at about 3.5 Tesla is due to the Landau level resonance condition between the upper and lower subband with $\Delta E_{21} = \hbar\omega_c$. Fig. 5.3(b) shows the current density versus magnetic field characteristics $J(B)$ at constant voltage. A clear increase in the current density is observed at the Landau level resonance due to elastic scattering between the upper and lower state of the optical transition. The possible elastic scattering mechanisms are interface roughness, ionized impurity, alloy disorder and electron-electron scattering. A so-called elastic current J_{elast} is extracted from the Landau level resonance, as shown in Fig. 5.3(c). The elastic current is related to the upper state population by

$$J_{elast} = \frac{J_{up}\tau_{up}}{\tau_{elast}} \tag{5.1.7}$$

τ_{up} is the total upper state lifetime and J_{up} is the current that is injected into the upper state, which is related to the total measured current density by the injection efficiency $J_{up} = \eta_{inj} J$. The ratio of the elastic to the total current density is then proportional to the injection efficiency

$$\frac{J_{elast}}{J} = \frac{\tau_{up}}{\tau_{elast}}\eta_{inj} \tag{5.1.8}$$

The elastic current can be extracted from magneto-transport measurements. For the relation 5.1.8 to be useful, it is important that the lifetimes τ_{up} and τ_{elast} are constant when the injection efficiency is non-zero. This condition can be considered as fulfilled. The elastic current is extracted along a Landau level resonance. The bias dependence of the upper and lower state of the radiative transition is negligible, once that the alignment regime is reached. Therefore the scattering mechanisms and lifetimes are approximately constant with the bias. The elastic current is extracted as indicated in Fig. 5.3(c) schematically. Fig. 5.3(d) shows an overview on the magneto-transport data. The second derivative of the current by the magnetic field d^2I/dB^2 allows to identify immediately the resonances through the black areas in the image plot. The lasing regime of the laser can be identified, since the threshold results in a change of the differential resistance. In conclusion, the relative injection efficiency into the upper state can be obtained by a simple analysis of the magneto-transport data.

5.1.4 Injection efficiency of low frequency QCLs

The technique described in the previous paragraph to deduce the relative injection efficiency is applied to the low frequency bound-to-continuum lasers. Three lasers have been investigated and compared, with frequencies of 1.7 THz (V303), 1.5 THz (N892) and 1.3 THz (N908). The relative injection efficiency for these three lasers as a function of the voltage is shown in Fig. 5.4(a,c,e). All three lasers show a similar behavior: the injection efficiency is zero below a certain voltage value, then it rises and reaches a maximal value close to the NDR voltage. At low bias, electrons are injected from the injector directly into the miniband, completely bypassing the

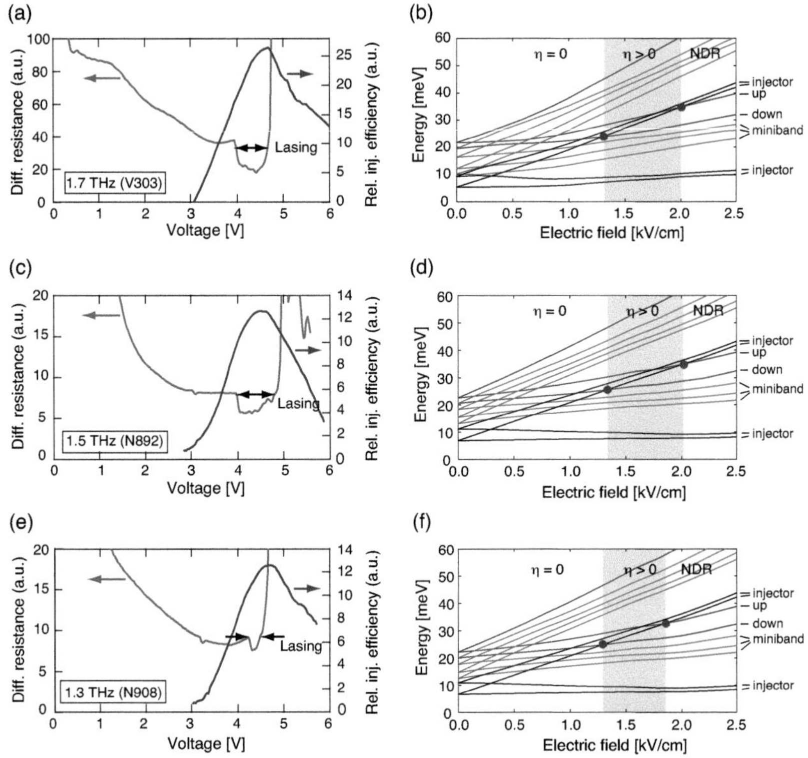

Figure 5.4: Relative injection efficiencies and corresponding simulation of the energy levels in the bandstructure of three lasers. The energy levels are obtained using one period and assuming weak coupling between the periods. At zero bias, the energy levels are degenerate, corresponding to two neighboring periods. The electric field lifts the degeneracy and rises the states of one period in respect to the other. The blue points indicate injection resonance into the highest level of the miniband and in the upper state.

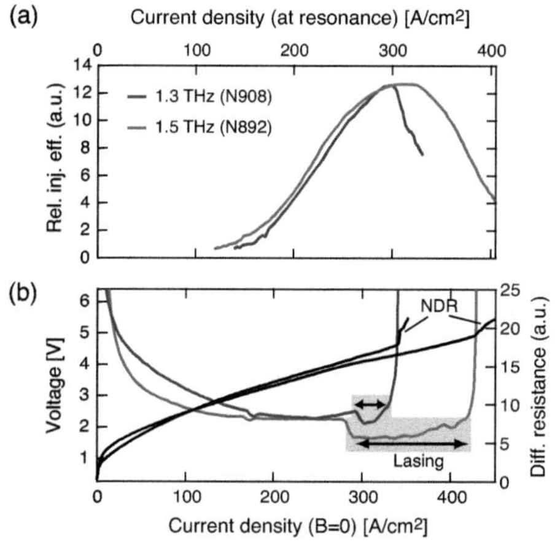

Figure 5.5: (a): Relative injection efficiency of the 1.5 THz and 1.3 THz laser as a function of the current density. (b): Properties of the two lasers without magnetic field.

upper state. The injection efficiency into the upper state starts after the bias that corresponds to the injection resonance of the injector state into the highest miniband state. The energy levels of the bandstructure as a function of the electric field are shown in Fig. 5.4(b,d,f). The injection resonance in the highest miniband state is marked with the first blue point. Beyond this bias, the injection efficiency increases and reaches a maximum at the injection resonance with the upper state (second blue point). Before the maximal injection efficiency is reached, electrons are injected in the miniband and represent an important leakage channel. This so-called wrong injection channel - from the injector into the lower state of the lasing transition - is very active at low bias, but loses its strength at the design electric field [47].

5.1.5 Comparison of the 1.5 and 1.3 THz laser

The bandstructure of the 1.5 THz and 1.3 THz laser differs only by a couple of monolayers in the first two quantum wells. As a consequence, the photon energy is smaller, but the maximal dynamic range is only 14% for the 1.3 THz laser compared to 32% for the 1.5 THz laser. A comparison of the relative injection efficiency as a function of the current density (along the Landau level resonance) is shown in Fig. 5.5. The regime of maximum injection efficiency is followed immediately by the NDR in the 1.3 THz laser, whereas the 1.5 THz laser remains electrically stable at the maximum injection efficiency and in addition, the NDR starts not exactly at maximum injection efficiency, but later. This difference might originate from the different coupling of the two injector states with the upper state as has been speculated in Sec. 4.4.2.

5.2 Towards a 1 THz laser

5.2.1 Scaling to 1.1 THz

The design of the 1.1 THz laser is based on a downscaled version of the 1.5 THz laser. A similar doping scheme is used to assure a proper alignment of the two injector states at injection resonance into the upper state. However the dopants are closer to the active region than in the 1.5 THz laser. The bandstructure is shown in Fig. 5.6(a) and the layer sequence is given in the figure caption. The dashed line indicates the doped area. At a field of 1.4 kV/cm the dipole is $z_{76} = 13$ nm and the transition energy is $E_{76} = 4.3$ meV. The period length is $L = 147.8$ nm.

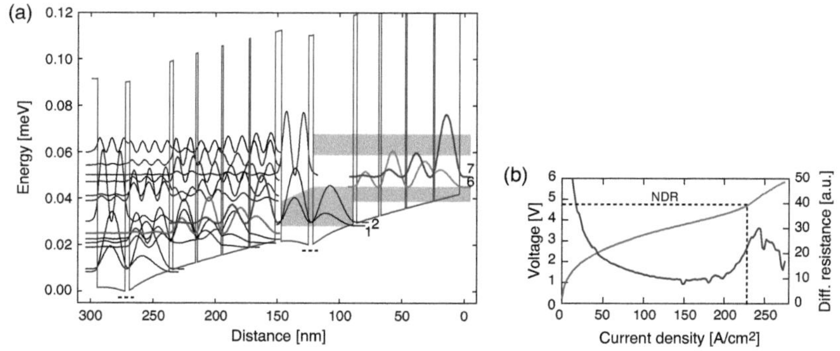

Figure 5.6: (a): Bandstructure of a 1.1 THz design. Self-consistent solution (T = 50 K) of coupled Schroedinger and Poisson's equations for an applied electric field of 1.4 kV/cm. The nominal layer sequence, starting from the injection barrier is (in nm): **4.9**/20.0/**1.0**/21.2/**1.3**/19.1/**1.8**/17.7/**2.9**/29.9/<u>2.3</u>/<u>**3.7**</u>/<u>2.2</u>/19.8. The figures in bold face represent the $Al_{0.1}Ga_{0.9}As$ barriers and the underlined layers are doped with Si, $5.0 \cdot 10^{16}$ cm^{-3} in density. The doping level yields a sheet carrier density of $4.1 \cdot 10^{10}$ cm^{-2} and an average doping density of $2.8 \cdot 10^{15}$ cm^{-3}. (b): Measured voltage versus current characteristics and the differential resistance. The onset of the NDR is indicated with a dashed line.

Processed samples have been measured, but no laser operation is observed. The electrical characteristics are shown in Fig. 5.6(b). An NDR region is identified at about 4.8 V.

A sample has been characterized in the magnetic field. The magneto-transport data is shown in Fig. 5.7(a). A transition with a strong Stark shift is identified through the Landau resonance corresponding to a cyclotron energy of $\sim$ 5 meV. Furthermore lasing action is identified through the change in the differential resistance at the threshold. The corresponding areas at about 4 T and from 7.5 -11 T are indicated in Fig. 5.7(a). Optical characterization at a field of 9 T is shown in Fig. 5.7(c,d). The laser threshold is at 5 V and lasing stops at 6 V. The corresponding spectra show emission at 1.1 THz. The relative injection efficiency indicates clearly that the injection into the upper state starts at 4 V with a maximum injection efficiency

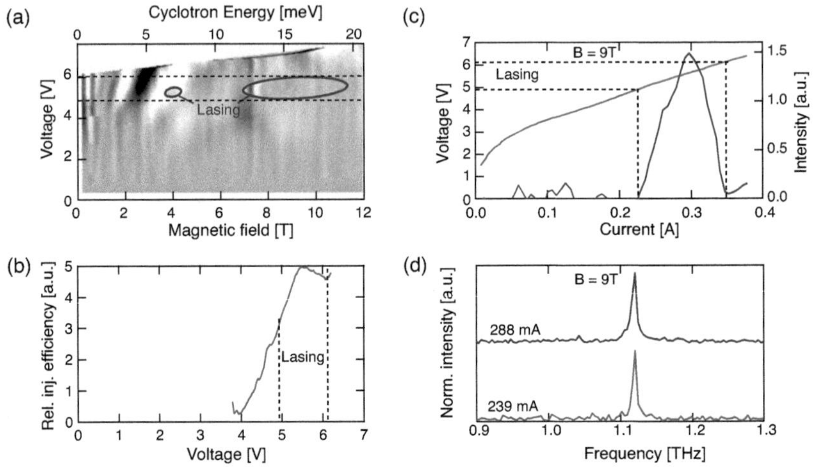

Figure 5.7: (a): Magneto-transport, showing the second derivation of the current by the magnetic field d^2I/dB^2. The blue areas indicate the lasing region. (b): The extracted relative injection efficiency. (c): The measured intensity versus the current at a field of 9T and (d): the measured spectra at different currents.

at about 5.5 V. The early NDR at 4.8 V is attributed to the mis-alignment of the injector with the miniband. The injection resonance of the injector with the upper state is at an electrically unstable working point and falls in the NDR region, due to the strong current channel prior to alignment. The current prior to alignment is quenched efficiently through the applied magnetic field, leading to an electrically stable operation at the injection resonance and hence to laser operation. The low field current maxima observed at about 4 V and 7 T might correspond to the Landau resonance of the miniband with the injector states.

The most likely explanation is that ionized impurity scattering leads to a broadening of the injector state and favors a broad non-selective injection. The strong current channel prior to alignment is responsible for the unstable working point at injection resonance.

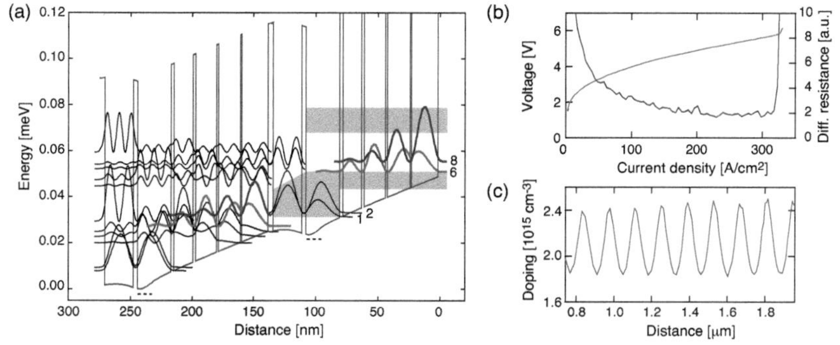

Figure 5.8: (a): Bandstructure of a 1.0 THz design. Self-consistent solution (T = 50 K) of coupled Schroedinger and Poisson's equations for an applied electric field of 1.8 kV/cm. The nominal layer sequence, starting from the injection barrier is (in nm): **4.0**/21.2/**1.0**/18.3/**1.4**/17.2/**2.0**/15.1/**2.5**/19.2/8.0/**3.3**/23.2. The figures in bold face represent the $Al_{0.1}Ga_{0.9}As$ barriers and the underlined layers are doped with Si, $6.3 \cdot 10^{16}$ cm^{-3} in density. The doping level yields a sheet carrier density of $5.0 \cdot 10^{10}$ cm^{-2} and an average doping density of $3.7 \cdot 10^{15}$ cm^{-3}. The nominal period length is 136.4 nm. (b): Measured voltage versus current characteristics and the differential resistance. (c): From capacitive measurements an average doping level of $2.2 \cdot 10^{15}$ cm^{-3} and a period length of 139.6 nm is deduced.

5.2.2 Scaling to 1.0 THz

The 1.0 THz design is again a scaled version of the 1.5 THz laser. The doping scheme is similar, and is used to assure a proper alignment of the two injector states at injection resonance into the upper state. The bandstructure is shown in Fig. 5.8(a) and the layer sequence is given in the figure caption. The dashed line indicates the doped area. At a field of 1.8 kV/cm the dipole is z_{86} = 11.1 nm and the transition energy is E_{86} = 4.5 meV. Processed samples have been measured, but no laser operation is observed. The electrical characteristics are shown in Fig. 5.8(b). The NDR occurs at 6 V. An average doping density of 2.2×10^{15} cm^{-3} is determined with a capacitive measurement. The measured doping density corresponds to 60% of the nominal doping.

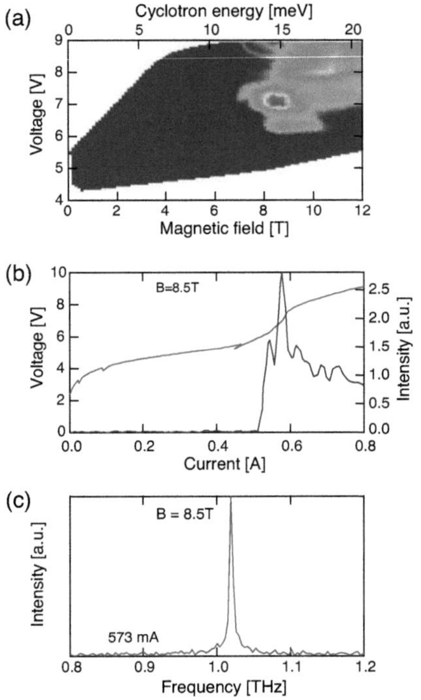

Figure 5.9: (a): Measured emission intensity as a function of the magnetic field and voltage. (b): Electrical characterization and measured intensity at $B = 8.5\text{T}$. (c): Measured spectrum at $B = 8.5\text{T}$.

Magneto-transport measurement of a sample has been carried out. Fig. 5.9(a) shows the measured emission intensity versus the magnetic field and voltage. Laser emission is observed at high magnetic fields, above 7 T. A clear emission maxima is observed at 8.5 Tesla, corresponding to a cyclotron energy of 14.7 meV. A Landau resonance in the injector well might lead to an efficient extraction of electrons from the lower state at this magnetic field value. Fig. 5.9(b) shows the voltage and intensity versus the current at a magnetic field of 8.5 T. The laser emission occurs in the NDR region, pointing out that the observed NDR corresponds to the mis-alignment of the parasitic current injection into the miniband. The magnetic field is quenching the non-radiative transitions; the NDR is less pronounced at 8.5 T than without magnetic field. Similar to the laser at 1.1 THz, the magnetic field allows to stabilize the bandstructure electrically at the design bias, through a decrease of the current prior to alignment into the miniband states. Laser emission close to 1.0 THz is observed at 8.5 T and shown in Fig. 5.9(c).

Again the 1.0 THz design suffers from the strong current channel prior to alignment. In addition the lower average doping could affect the proper alignment of the injector states leading to a less efficient injection into the upper state.

5.2.3 4 well laser at 1.2 THz

The bandstructure in Fig. 5.10(a) is designed with the objective to increase the injection efficiency into the upper state and stabilize the structure at the design bias. For this reason the injector doublet has been reduced to one injector state [106]. Furthermore the active region has been shortened by one well leading to a miniband with three instead of four states. The resulting bandstructure reassembles

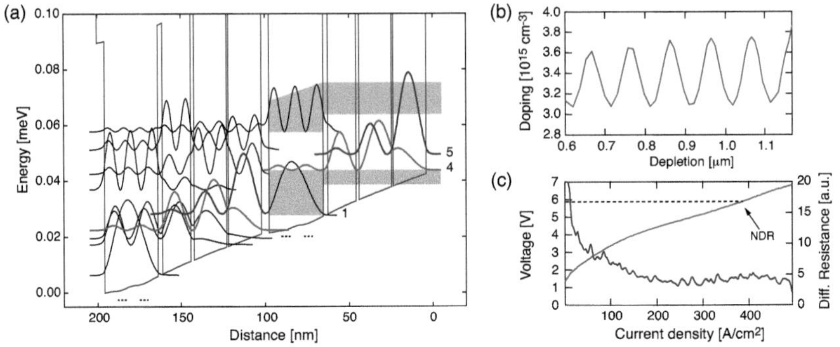

Figure 5.10: (a): Bandstructure of a 1.2 THz design. Self-consistent solution (T = 50 K) of coupled Schroedinger and Poisson's equations for an applied electric field of 1.8 kV/cm. The nominal layer sequence, starting from the injection barrier is (in nm): **4.6**/19.6/**1.0**/19.4/**1.8**/17.0/**2.8**/65/60/70/60/65. The figures in bold face represent the $Al_{0.1}Ga_{0.9}As$ barriers and the underlined layers are doped with Si, $5.0 \cdot 10^{16}$ cm^{-3} in density. The doping level yields a sheet carrier density of $6.0 \cdot 10^{10}$ cm^{-2} and an average doping density of $6.1 \cdot 10^{15}$ cm^{-3}. (b): From capacitive measurements an average doping level of $3.3 \cdot 10^{15}$ cm^{-3} and a period length of 99.7 nm is deduced. (c): Measured voltage versus current characteristics and the differential resistance.

to a scaled version of the 4 well laser - but without the phonon extraction resonance - [140]. The injector well is doped symmetrically, avoiding the center of the well and the region close to the injector barrier. At the alignment bias, the computed photon energy is E_{54} = 5.6 meV, and the dipole is z = 10.1 nm. From capacitive measurements an average doping density of 3.3×10^{15} cm^{-3} is deduced, that is about 54% of the nominal doping. No laser operation is observed without magnetic field. The measured voltage versus current characteristics and the differential resistance are shown in Fig. 5.10(c). A weak NDR region is identified at a bias of 6 V.

Magneto-transport studies are carried out on a sample of this layer. Fig. 5.11(a) shows an overview. Lasing operation is achieved above a magnetic field of 6 T. The relative injection efficiency is deduced from the Landau resonance close to 5 meV between the upper and lower lasing level, shown in Fig. 5.11(b). The injection

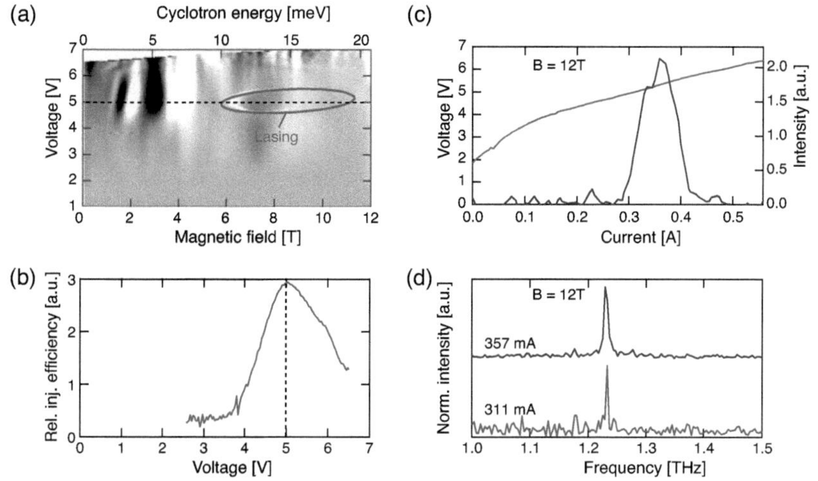

Figure 5.11: (a): Magneto-transport showing the second derivative of the current by the magnetic field d^2I/dB^2. The lasing area is indicated by a red oval. (b): The relative injection efficiency extracted along the Landau level resonance at about 5 meV. (c): Optical characterization at a magnetic field of 12 T. (d): Spectra measured at 12 T.

efficiency is maximal at 5 V, fully in agreement with the maximal emission measured at a voltage of 5 V at 12 T in Fig. 5.11(c). The lasing frequency is close to 1.2 THz, shown in Fig. 5.11(d). In contrast to the 1.1 THz and 1.0 THz laser, this laser is stable at the design bias. However the gain without magnetic field is not high enough to reach threshold. The lower doping level shouldn't affect the proper alignment of the bandstructure, however it lowers the maximal achievable gain in the structure. The lifetime of the lower state might be longer, since the scattering phase space is reduced, compared to the other bound-to-continuum structures with four miniband states and the extraction resonance could be a bottle-neck in this structure. An indication for this hypothesis is that the maximal gain is achieved prior to injection resonance and the onset of the NDR. The lower state lifetime could be affected by the alignment of the miniband states with the excited state of the injector well.

5.3 Conclusions

The relative injection efficiency is extracted from magneto-transport measurements of low frequency terahertz QCLs. Three laser designs aiming the frequency range below 1.2 THz have been investigated. They do not show laser operation without magnetic field, but in the magnetic field laser operation is observed. The limiting mechanisms are studied through magneto-transport measurements. The 1.1 THz and 1.0 THz laser are based on a downscaled bound-to-continuum design with split injector. For both lasers, the alignment bias corresponds to an electrically unstable bias due to the strong current channel prior to alignment. Ionized doping scattering of the injector state is suspected for non-selective injection. The 1.2 THz laser is based on a four well design. The structure is electrically stable at the alignment bias. However the population inversion is not strong enough. A possible explanation is a longer lower state lifetime due to the reduced scattering phase space of the miniband.

Future designs based on the bound-to-continuum design with split injector should address the doping issue. A doping profile similar to the lasers operating above 1.6 THz should be re-adopted. To increase the injection selectivity, thicker injector barriers could be considered as well. On the other hand, an optimization of the four well design, in particular the extraction resonance, might lead to low frequency operation with this design.

Chapter 6

The lumped circuit Laser

6.1 Introduction

6.1.1 Microcavity lasers

An optical microcavity is characterized by its ability to confine an optical mode into a small volume. A key number is the effective mode volume V_{eff} taken by the energy density of the oscillating electric field. The normalized mode volume for various micorcavity lasers is shown in Fig. 6.1.

The minimal mode volume that can be achieved based on dielectric confinement is in the order of $\sim 2(\lambda/2n_{eff})^3$, where λ is the wavelength, and n_{eff} is the effective refractive index [149]. Microcavity lasers with mode volumes close to this limit have been demonstrated. At telecom wavelength ($\lambda = 1.5\mu$m) an optically pumped microcavity laser formed from a single defect in a two-dimensional photonic crystal with a mode volume of only $2.5(\lambda/2n_{eff})^3$ and a quality factor $Q = 250$ has been reported [145]. An electrically driven single mode photonic band gap laser has been

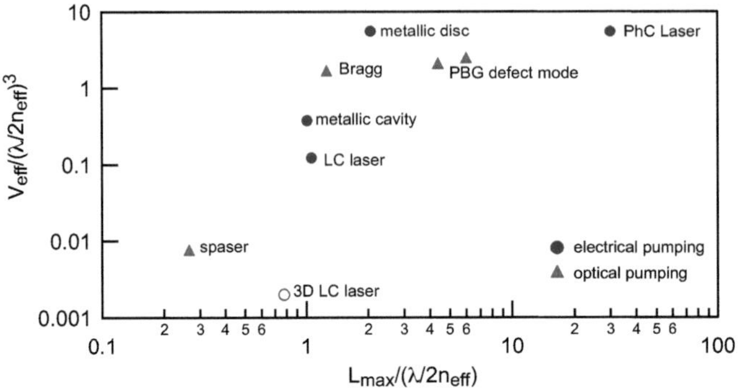

Figure 6.1: Normalized effective mode volume as a function of the largest normalized mode diameter for various microcavity lasers based on electrical or optical pumping as indicated. The mode volume is estimated, if not specified explicitly, from the published data. Metallic disc [141], PhC laser [142], Bragg [143], PBG defect mode [144, 145], metallic cavity [146], spaser [147], LC laser [148]. Empty symbol, 3D LC laser: in principle very low mode volumes could be achieved with the LC laser, as will be discussed in Sec. 6.5.1.

demonstrated as well, with a mode volume of $5.5(\lambda/2n_{eff})^3$ and a quality factor of $Q > 2500$ [142].

Smaller mode volumes are achieved with metallic confinement. Metallic boundaries prevent out-leaking of modes that are subject to a sub-wavelength confinement. Electrically pumped, metallic disc lasers have been demonstrated in the terahertz range with mode volumes of $\geq 5.6(\lambda/2n_{eff})^3$ and a typical quality factor of 80 [141, 150, 151]. At $\lambda = 1.4\mu$m, an electrically pumped laser has been reported, based on a cylindrical metallic cavity that has a mode volume of $0.38(\lambda/2n_{eff})^3$, well below the limit of dielectric confinement and a quality factor $Q > 140$ [146]. Even higher confinement and smaller mode volumes are achieved by exploiting explicitly the plasmonic nature of metals. A strong sub-wavelength confinement in two dimensions has been realized, using an optically pumped nanowire close to a silver surface to make use of the confinement by surface plasmons. The laser has a

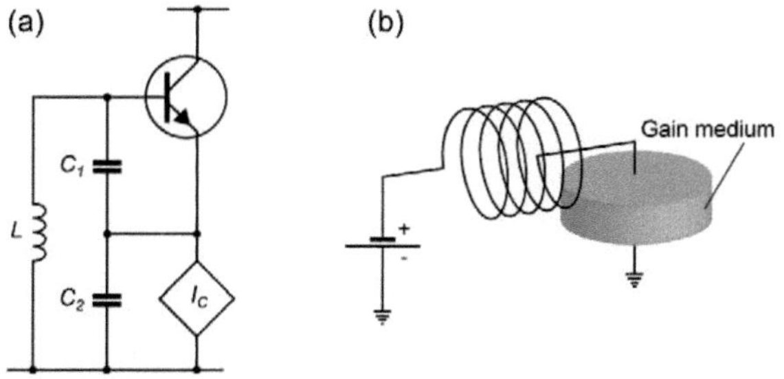

Figure 6.2: (a): Electronic oscillator based on current amplification with a transistor. (b): LC laser-oscillator, based on amplification of the electric field in an optical gain medium.

quality factor of about 27 and is lasing at $\lambda = 490$ nm [152]. A gold nanoparticle has been used as a nano-antenna for the enhancement of the fluorescence emission of a single molecule [153]. Recently a group has claimed the demonstration of an optically pumped nanolaser based on the plasmonic nanoparticle resonance of a gold nanoparticle embedded in a dye-doped silca shell. A mode volume of about $0.008(\lambda/2n_{eff})^3$ and a quality factor of 15 at an emission wavelength of $\lambda = 530$ nm is reported [147].

The LC laser uses a circuit approach to obtain a sub-wavelength mode confinement [148]. LC laser stands for inductor-capacitor laser. Inspired from electronics, a resonant lumped circuit resonator comprising an inductor and a capacitor is designed at terahertz frequencies. The active medium of a QCL is embedded between the capacitor plates. Fig. 6.2 shows the comparison of an electronic oscillator and the principle of the LC laser. Both are based on a lumped circuit inductor-capacitor resonator. But while a transistor is used to amplify the current in the electronic oscillator, the electric field between the capacitor plates in the LC laser is amplified by stimulated emission of radiation. By concept the LC laser is sub-wavelength

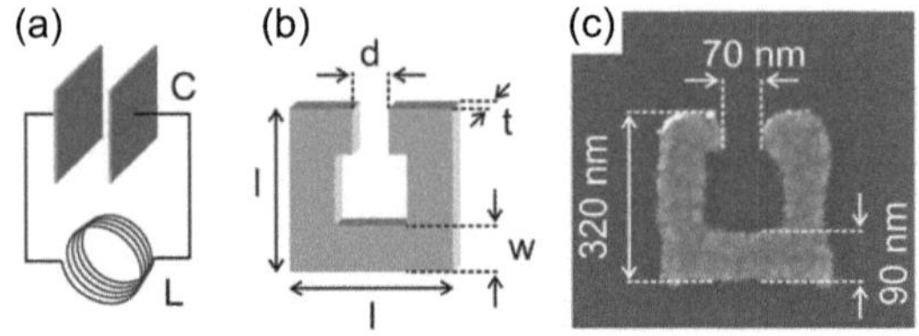

Figure 6.3: (Figure and description from Ref. [155]) Demonstration of a magnetic metamaterial at 100 THz. (a): Illustration of the analogy between a conventional LC circuit and (b): the single SRRs used. (c): An electron micrograph of a typical fabricated SRR.

sized and should be scalable at will in principle. The circuit approach in optics is not new; the lumped circuit element approach at optical frequencies was proposed by Engheta [154] and passive resonant electronic circuits are commonly used in metamaterials.

6.1.2 Metamaterials

From the electromagnetic point of view, the wavelength λ, determines whether a collection of atoms or other objects can be considered as an effective material. The effective electromagnetic parameters ε and μ need not to arise strictly from the response of atoms or molecules. Any collection of objects whose size and spacing are much smaller than λ can be described by an effective ε and μ [156]. In metamaterials with simultaneous negative ε and μ the group and phase velocities point in opposite directions, therefore called negative index material. They don't exist naturally. Their properties have been investigated theoretically [157] a long time before their experimental demonstration [158].

Metamaterials from the microwave to the visible spectrum have been realized by building large arrays of a unit cell consisting of a sub-wavelength sized metallic structure. A frequently used structure is the split ring resonator (SRR) [159]. Metamaterials with magnetic and electric response in the terahertz range [160, 161] and even up to the near-infrared range [155] have been demonstrated. An example of a sub-wavelength planar cell that is used for magnetic response at 100 THz is shown in Fig. 6.3. The metallic structure behaves as a resonant electronic circuit. Similar, the concept of the LC laser relies on a resonant electronic circuit at optical frequencies, but in contrast to metamaterials it is an active device. The LC laser can be seen as implementing gain in one cell of a LC circuit based metamaterial [162].

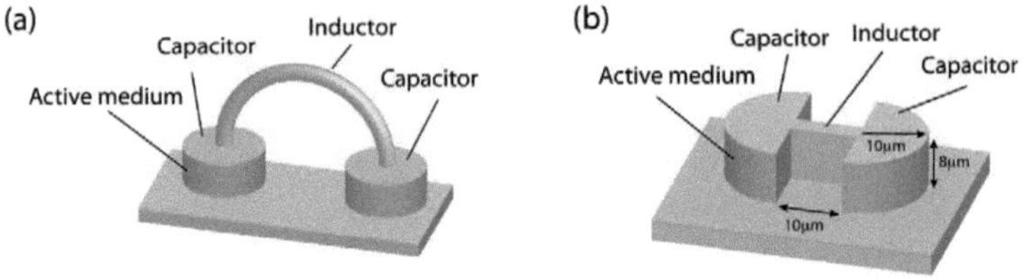

Figure 6.4: (a): The ideal LC resonator, consisting of two capacitors and a wire inductor. (b): The chosen resonator geometry for the implementation of the LC resonator. A planar inductor allows easier fabrication.

6.2 LC resonator

6.2.1 From the ideal to the real device

An ideal LC resonator that fits the requirements for the implementation of a gain medium between the capacitor plates is shown in figure 6.4(a). It consists of two capacitors and an inductor wire that connects the two capacitors. Two capacitors are required since the active region between the capacitor plates is pumped electrically. In the simplest scheme, with only one capacitor and one inductor, the pump bias would be shorted. A terahertz QCL active gain medium is inserted between the capacitor plates. The electric field between the capacitor plates is perpendicular to the layers in the active region that define the quantum wells and therefore couples to the intersubband transition. The target frequency for the resonator is 1.5 THz, corresponding to a free space wavelength of 200 μm.

An LC resonator is by concept sub-wavelength sized. Fabrication of such a device with a free standing wire is possible (using direct laser writing [163]), but not a trivial task. For a proof of principle, a simpler geometry is preferred as shown in Fig. 6.4(b). A planar implementation of the inductor is chosen. The capacitors are half-circular shaped. The precise shape of the capacitors is not relevant. The half-circular shape has been chosen for fabrication reasons and to have an high mode overlap with the active region. Gold is chosen for the metallic parts. The length of the resonator is 30 μm and the thickness of the active region is 8 μm. The dielectric around the resonator is removed by etching. Although an etching step is not absolutely required, it is a preferred situation since it increases the confinement of the optical mode to the active region.

The resonator is characterized by the resonance frequency, the quality factor due to ohmic and radiative losses, the mode volume and the overlap factor of the mode with the active region. Three models are considered with different degrees of complexity to address the relevant parameters of the resonator. With a simple model the resonance frequency is estimated. A 3D fullwave simulation allows to compute the fields and all relevant parameters, on the expense of time consuming simulations and demanding computation powers. Finally, an analytic microwave model is developed, for a fast and efficient computation of the resonance frequency and the ohmic losses.

6.2.2 Simple model of the LC resonator

The scope of the simple model is to estimate the resonance frequency of the LC resonator. The electrical values of the capacitor and inductor are computed with simple well-known formulas. The half-circular shaped capacitors are treated as ideal planar capacitors, applying the formula

$$C = \varepsilon_0 \varepsilon \frac{A}{d}$$

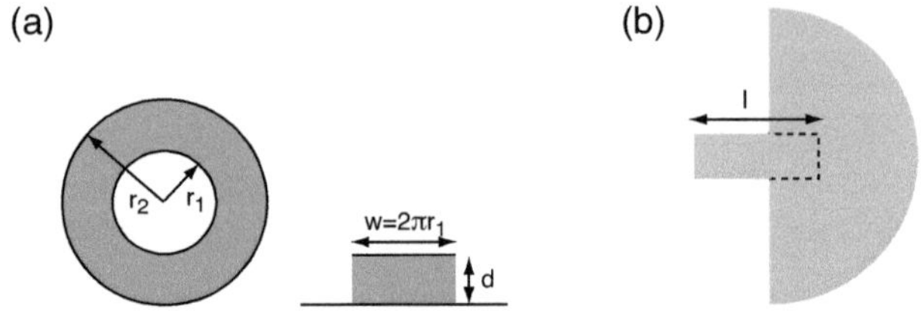

Figure 6.5: (a) A section through a coaxial cable and a section through the inductor of the LC resonator. Identification of r_1 and r_2 allows to consider the inductor of the LC resonator as an unrolled coaxial cable. (b): The inductor length is measured to the center of mass of the capacitor plate.

where A is the surface of the capacitor plate and d the distance between the plates. ε is the relative dielectric constant of the semiconductor sandwiched between the plates. For Gallium Arsenide a value of $\varepsilon = 12.96$ is used.

A rough estimate of the inductance is obtained when considering the inductor wire as an unrolled coaxial cable as shown in Fig. 6.5(a). The formula for the inductance of a coaxial cable is

$$L = \frac{\mu_0 l}{2\pi} \ln\left(\frac{r_2}{r_1}\right)$$

where r_2 is the outer and r_1 is the inner radius of the coaxial cable, and l is the length of the inductor, measured to the center of mass of the capacitor plate [Fig. 6.5(b)]. The planar inductor's width w is related to r_1 and r_2 by: $w = 2\pi r_1$ and $r_2 = r_1 + d$, where d is the thickness of the active region.

The resonance frequency of the resonator is obtained by the formula for a LC circuit

$$f = \frac{1}{2\pi\sqrt{LC}}$$

The symmetry requires to compute only the L and C values of the half resonator, since the full resonator has then twice the inductance but half the capacitance, leading to the same resonance frequency. The equivalent L and C values and the resonance frequency of the LC resonator in Fig. 6.4(b) are $L = 4.79$ pH, $C = 2.25$ fF and $f = 1.53$ THz.

6.2.3 Fullwave simulation

3D fullwave simulations of the LC resonator are carried out using the commercial software package Comsol [164]. Comsol allows to model and solve problems that are based on partial differential equations, such as electromagnetic problems. To solve numerically the partial differential equations, Comsol uses the finite element method, where the volume is divided into subdomains, called finite elements. The software runs the finite element analysis together with adaptive meshing and error control using a variety of numerical solvers.

Eigenmode analysis and harmonic propagation are used to compute the resonance frequency af the LC resonator and the losses. Two loss mechnisms are identified; ohmic losses in the metal layers and radiative losses. The losses are quantified by their respective quality factors Q. The Q is related to the photon lifetime τ_p by $Q = \omega\tau_p$ and to the losses by $Q = \omega/(\alpha\tilde{c})$, where $\tilde{c}$ is the velocity of light in the material.

Resonance frequency

The resonance frequency of the LC resonator is efficiently and accurately calculated using eigenmode analysis. The LC resonator is simulated with the ground plane,

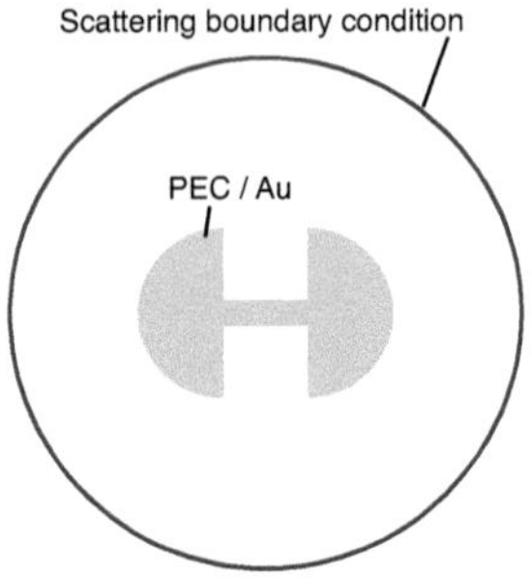

Figure 6.6: Eigenvalue simulation to determine the resonance frequency of the LC resonator. If the metal layers of the LC resonator are modeled using the optical parameters of the metal, the ohmic Q can be determined.

and embedded into a spherical space limited by scattering boundary conditions. Since the mode is strongly confined to the resonator, the boundary conditions are not affecting the resonance frequency of the mode. The metallic layers of the LC resonator are modeled as perfect electrical conductors (PECs), which can also be defined as internal boundary conditions in the model. Fig. 6.6 shows a schematic of the simulation. The eigenfrequency, electric and magnetic fields are computed fast and precise in about 1 minute using a computer with a modern processor.

Ohmic losses

Ohmic losses are computed carrying out again an eigenmode analysis, but the metal layers are modeled as a real metal, using the complex refractive index of gold (Au) at the resonance frequency of the resonator. The eigenfrequency analysis solves for the eigenfrequency of the LC resonator. Since the resonator involves losses, the time-harmonic representation of the fields includes a complex parameter in the phase

$$\mathbf{E}(\mathbf{r},t) = \mathcal{R}e[\tilde{\mathbf{E}}(\mathbf{r})e^{-\lambda t}] \tag{6.2.1}$$

where the eigenvalue is $(-\lambda) = \delta + i\omega$. The imaginary part represents the eigenfrequency, and the real part is responsible for the damping. The quality factor Q is derived from the eigenfrequency and damping

$$Q = \frac{\omega}{2|\delta|} \tag{6.2.2}$$

Since the scattering boundary condition is used, part of the radiation leaves the resonator, adding additional losses (Fig. 6.6). Therefore the simulation is run twice, once with the metal layers modeled as perfect electrical conductors to determine the Q corresponding to the radiation losses through the outher boundaries, then the simulation is run again with the metal layers modeled as gold. A comparison allows to determine the net ohmic Q. The simulation using the optical paramters of gold is very demanding. The skin depth in gold at 1.5 THz is about 50 nm, requiring a very fine meshing of the gold layers that results in a large required memory and a long computation time. A simulation limited by 16 GB of memory takes 1-2 hours. For the complex refractive index of gold $\hat{n} = n - ik$, the values $n = 431$ and $k = 543$ are used at 1.5 THz. These values are determined with the Drude model based on the parameters of reference [116]. Comsol uses the convention of a $e^{i\omega t}$ dependence of the time harmonic fields, therefore the minus sign in the imaginary part of the refractive index is required.

Radiative losses

One of the challenges in finite element modeling is how to treat open boundaries in radiation problems. Two closely related types of absorbing boundary conditions are offered in Comsol, the scattering boundary condition and the matched boundary condition. The former is perfectly absorbing for a plane wave, whereas the latter is perfectly absorbing for guided modes, provided that the correct value of the propagation constant is supplied. However, in many antenna-modeling problems, the incident radiation cannot be described as a plane wave with a well-known direction of propagation. In such situations, perfectly matched layers (PMLs) are useful. A perfectly matched layer is strictly speaking not a boundary condition but an additional domain that absorbs the incident radiation without producing reflections. It provides good performance for a wide range of incidence angles and is not particularly sensitive to the shape of the wave fronts. The perfectly matched layer formulation can be deduced from Maxwell equations by introducing a complex-valued coordinate transformation under the additional requirement that the wave impedance should remain unaffected [165].

The harmonic propagation analysis in Comsol supports the direct computation of the S-parameters and far-field patterns. Perfectly matched layers can be included in the simulation and the excitation is defined through a port. Comsol allows in particular to define lumped ports, that are useful for antenna simulations, but also for the simulation of the radiation properties of the LC resonator. The lumped port is applied between two metallic objects separated by a distance much less than the wavelength. Comsol calculates the impedance Z_{port} for the lumped port. The impedance is directly given by the relation

$$Z_{port} = \frac{V_{port}}{I_{port}} \tag{6.2.3}$$

where V_{port} is the extracted voltage for the port given by the line integral of the electric field between the terminals averaged over the entire port. The current I_{port} is the averaged total current over all cross sections parallel to the terminals. The Q of a wide variety of antennas has been shown to be accurately approximated from the antenna's input impedance under the condition of a single sufficiently isolated resonance with $Q \gg 1$. These conditions are satisfied in the LC resonator. The Q is given by [166, 167]

$$Q_{rad} = \frac{\omega_0}{2R_0} |Z_0'(\omega_0)| \tag{6.2.4}$$

where ω_0 is the angular frequency at which the antenna is tuned to zero reactance, R_0 is the resistance of the antenna at ω_0 and $Z_0'(\omega_0)$ is the frequency derivative of the antenna impedance at ω_0. Eq. 6.2.4 can be considered as the Q of an equivalent series RLC circuit, where the R, L, and C values are determined from the computed impedance at resonance $Z(\omega_0)$. To show this, a series RLC circuit is considered. The impedance of the circuit is

$$Z = R + i\omega L + \frac{1}{i\omega C} \tag{6.2.5}$$

and the corresponding quality factor is expressed as $Q = \omega L/R$. The derivative of Eq. 6.2.5, evaluated at the resonance frequency ω_0 is $|Z'(\omega_0)| = 2L$, and Eq. 6.2.4

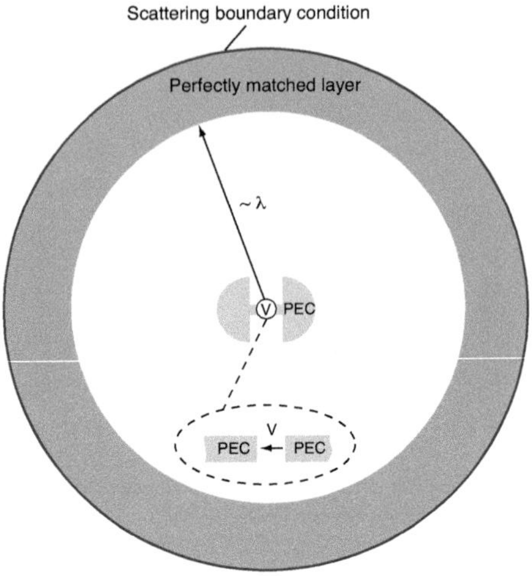

Figure 6.7: The computation of the radiative Q is based on a harmonic propagation analysis with a lumped port for excitation. Open boundaries are modeled through a perfectly matched layer and scattering boundary conditions.

is recovered.

For the simulation of the radiative losses, the metallic layers of the LC resonator are modeled as perfect electrical conductors, therefore the only loss mechanisme are radiation losses. The perfectly matched layer combined with scattering boundary conditions to minimize reflection, is defined at a distance of about a wavelength from the resonator, resulting in a large volume to be meshed. The lumped port is defined at the symmetry point of the resonator, as shown in Fig. 6.7. The harmonic propagation analysis requires a series of simulations where the frequency is the parameter. A well resolved series of simulations requires computation times in the order of an hour.

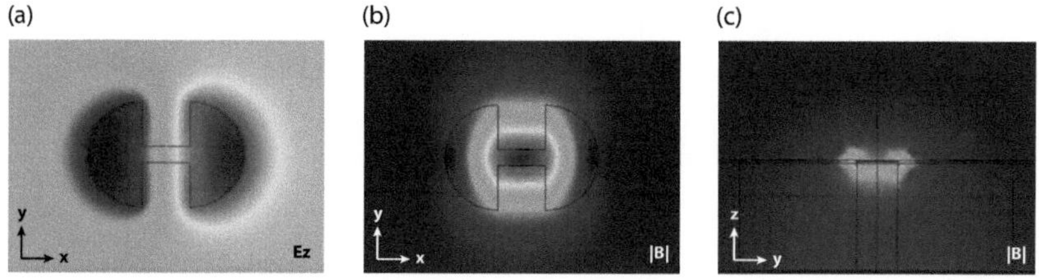

Figure 6.8: Fullwave simulation of the electromagnetic field in the resonator. (a): The dominant electric field component E_z between the capacitor plates and (b): the magnitude of the magnetic field $|\vec{B}|$ at mid-resonator height. (c): The magnetic field magnitude on a section through the inductor.

Results

Fig. 6.8 shows the fields of an eigenmode analysis, the gold being modeled with a complex refractive index. The electric field is mainly polarized in the z direction, perpendicular to the capacitors plates, as required by the selection rules of the active gain material. A plot of the electric field component E_z in the resonant mode on a section through the LC resonator between the capacitor plates in Fig. 6.8(a) shows that the electric field is antisymmetric with respect to the axis of symmetry of the structures and is mostly concentrated and uniform under the capacitor plates. The magnetic field in Fig. 6.8(b,c), on the other hand, is mainly concentrated around the inductor. The fields confirm the interpretation in terms of a capacitor-inductor-capacitor resonant circuit. Furthermore a confinement factor of the optical mode of $\Gamma = 0.85$ and a resonance frequency of 1.45 THz are computed. The quality factor due to ohmic losses caused by conduction currents in the gold layers is $Q_{ohm} = 53$ and the radiative quality factor is $Q_{rad} = 189$. The resonator is expected to exhibit a final $Q = 41$, limited by the ohmic losses.

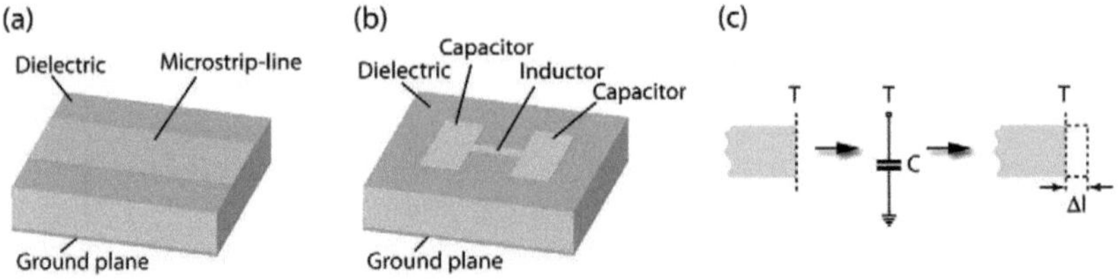

Figure 6.9: (a): Geometry of a microstrip-line. (b): The LC resonator is modeled as a quasi-lumped resonator. (c): An open end microstrip-line is modeled as an equivalent shunt capacitor or with an equivalent length of the transmission line Δl.

6.2.4 Microwave model

The resonance frequency and ohmic losses of the LC resonator can be computed using analytical formulas that are based on a microwave model. The benefits of analytical expressions is that they allow a fast and efficient modeling, since the calculation time is in the order of a second.

At microwave frequency, microstrip-lines are commonly used for the design of radio frequency and microwave applications such as filters or antennas [117, 168]. The microstrip-line consists of a conducting strip separated by a dielectric material from a ground plane, shown in Fig. 6.9(a). Microstrip-lines are transmission lines. However stubs and sections that are shorter than 1/4 of the guided wavelength act as quasi lumped elements. Depending on their geometry they behave rather as a capacitor or inductor, and are described by an equivalent electronic circuit.

The LC resonator of Fig. 6.4(b) is modeled as three connected short sections of microstrip-lines, shown in Fig. 6.9(b). To keep the model simple, the capacitor is modeled with a rectangular instead of a half-circular shape, but with the same area. This simplification has little influence, since to a first order the precise shape of the capacitor plates doesn't matter.

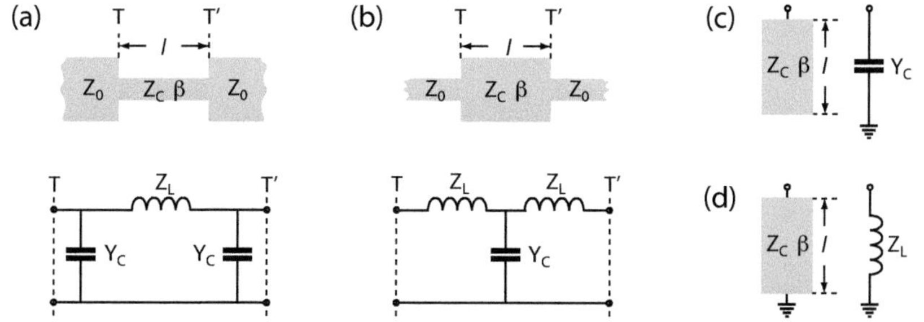

Figure 6.10: Quasi-lumped elements are shorter than 1/4 of the guided wavelength. Lossless (a): high-impedance short-line element and the equivalent circuit, (b): low-impedance short-line element with the equivalent circuit, (c): open circuited stub and the equivalent capacitor, (d): short-circuited stub and the equivalent inductor.

Fullwave simulations and experiments have confirmed that the resonance frequency of etched resonators is substantially higher [typically 30% for the LC resonator of Fig. 6.4(b)] than for resonators with an un-etched substrate. The main influence of un-etched substrates comes from the electric field leaking into the substrate at the open end of both capacitors, leading to effectively larger capacitors for the un-etched resonator. To take this effect into account, an open end of a microstrip-line with an un-etched substrate is modeled as a shunt capacitor or equivalently as a longer effective length, shown in 6.9(c). The formula for the correction of the effective length is given in the appendix Sec. B.2. The dielectric at the lateral boundaries of the microstrip-lines has little influence on the propagation constant, especially for thin substrates, therefore it doesn't require a correction. The calculation of etched resonators doesn't need the correction for the effective length.

Equivalent circuit of quasi-lumped elements

Microstrip-lines are described by the characteristic impedance Z_c and the effective dielectric constant ε_{re}, that depend on the geometry of the microstrip-line. Analytical expression are given in Appendix B.1. The complex propagation constant is computed as

$$\gamma = \alpha + i\beta \tag{6.2.6}$$

The units of the propagation constant are $[\gamma]$=1/length. The losses α are defined by the amplitude attenuation (for optical resonators, the losses are often defined by the power attenuation) and $\beta = 2\pi\sqrt{\varepsilon_{re}}/\lambda_0$, where λ_0 is the free space wavelength.

Since for the microstrip-lines considered here, $\alpha \ll \beta$ by at least an order of magnitude, the propagation constant is mainly determined by β. Therefore quasi-lumped elements are first considered for the case where $\alpha = 0$. Fig. 6.10(a) shows a short section of an high-impedance microstrip-line terminated at both ends by relatively low impedance and the equivalent circuit. The dual case is shown in Fig. 6.10(b), a short section of low-impedance microstrip-line is terminated at either end by relatively high impedance. Open respectively short-circuited stubs are equivalent to a capacitor respectively an inductor as shown in Fig. 6.10(c) and (d). The equivalent circuit parameters are given by the following expressions, where Z is the impedance and $Y = 1/Z$ is the admittance.

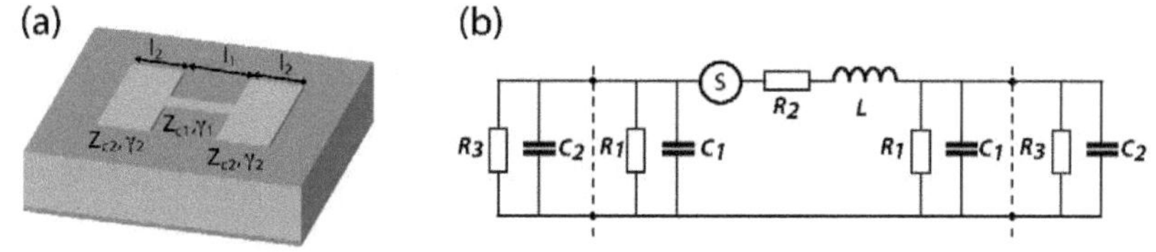

Figure 6.11: (a): Quasi lumped resonator, and (b): equivalent circuit.

High Z short-line	Low Z short-line	Open-circuited stub	Short-circuited stub
$Z_L = iZ_c \sin(\beta l)$	$Z_L = iZ_c \tan(\beta l/2)$	$Y_C = iZ_c^{-1} \tan(\beta l)$	$Z_L = iZ_c \tan \beta l$
$Y_C = iZ_c^{-1} \tan(\beta l/2)$	$Y_C = iZ_c^{-1} \sin(\beta l)$		

Z_L respectively Y_C are the impedances respectively admittances defined in the circuits of Fig. 6.10. Z_c is the characteristic impedance, and β the imaginary part of the propagation constant of the corresponding microstrip-line.

If losses are not neglected, the above stated formulas are still valid if $i\beta$ is substituted by the propagation constant γ. The impedances are no longer purely imaginary. The real part of the impedances corresponds to resistors, that take into account the ohmic losses. Since $\alpha l \ll \beta l < \pi/2$, the trigonometric terms in αl can be expanded to the first order. The resulting impedance corresponds to an equivalent circuit with identical capacitors and inductors, but the inductors have an additional series resistor and the capacitors an additional parallel resistor. The possibility to include the losses in terms of resistors in the calculation, allows to determine the ohmic Q of the circuit.

Equivalent circuit of the LC resonator

The different sections of the LC resonator shown in Fig. 6.11(a) are computed as a high-impedance short-line and open circuited stubs on either side. The equivalent circuit is shown in Fig. 6.11(b). Following expressions for the electrical components are obtained using the formulas of the previous paragraph. The exact expressions and a first oder expansion in αl are

$$\begin{aligned}
\omega L &= Z_{c1}\cosh(\alpha_1 l_1)\sin(\beta_1 l_1) \approx Z_{c1}\sin(\beta_1 l_1) \\
R_2 &= Z_{c1}\sinh(\alpha_1 l_1)\cos(\beta_1 l_1) \approx Z_{c1}\alpha_1 l_1\cos(\beta_1 l_1) \\
\omega C_1 &= Z_{c1}^{-1}\frac{\tan(\beta_1 l_1/2)}{\cosh^2(\alpha_1 l_1/2)+\sinh^2(\alpha_1 l_1/2)\tan^2(\beta_1 l_1/2)} \approx Z_{c1}^{-1}\tan(\beta_1 l_1/2) \\
1/R_1 &= Z_{c1}^{-1}\frac{\tanh(\alpha_1 l_1/2)}{\cos^2(\beta_1 l_1/2)+\sin^2(\beta_1 l_1/2)\tanh^2(\alpha_1 l_1/2)} \approx Z_{c1}^{-1}\frac{\alpha_1 l_1/2}{\cos^2(\beta_1 l_1/2)} \\
\omega C_2 &= Z_{c2}^{-1}\frac{\tan(\beta_2 l_2)}{\cosh^2(\alpha_2 l_2)+\sinh^2(\alpha_2 l_2)\tan^2(\beta_2 l_2)} \approx Z_{c2}^{-1}\tan(\beta_2 l_2) \\
1/R_3 &= Z_{c2}^{-1}\frac{\tanh(\alpha_2 l_2)}{\cos^2(\beta_2 l_2)+\sin^2(\beta_2 l_2)\tanh^2(\alpha_2 l_2)} \approx Z_{c2}^{-1}\frac{\alpha_2 l_2}{\cos^2(\beta_2 l_2)}
\end{aligned}$$

The equivalent electric components, e.g. capacitors, inductors and resistors have frequency dependent values. If the length of the microstrip-line sections and stubs were much smaller than the guided wavelength, i.e. $l < \lambda_g/10$ then a first order approximation of the terms in βl would lead to frequency independent values for capacitors and inductors.

The frequency behavior of the circuit is analyzed by computing the impedance $Z(\omega)$ seen by the excitation (source S), placed in the symmetry point of the resonator, in the middle of the high-impedance short-line section. The resonance condition of the

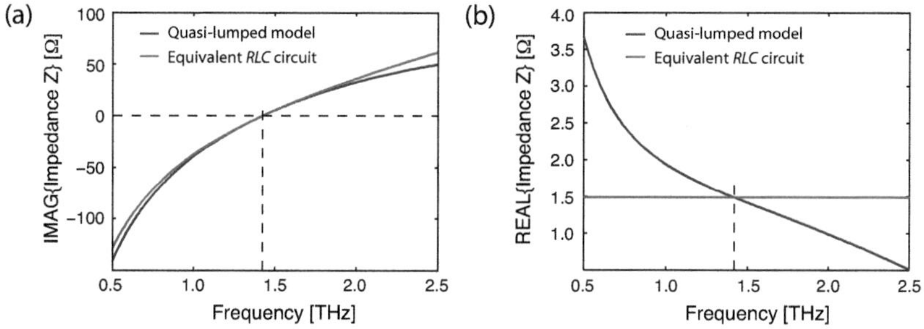

Figure 6.12: Comparison of the (a): imaginary and (b): real part of the impedance computed as a quasi-lumped resonator to the impedance of the equivalent *RLC* circuit. The resonance condition is reached when the imaginary part of the impedance goes to zero.

circuit is reached when the imaginary part of the impedance goes to zero. Although the electric components have frequency dependent values, an equivalent series *RLC* circuit can be determined that reproduces with high accuracy the impedance of the circuit of Fig. 6.11(b). The impedance of a series *RLC* circuit is

$$Z = R + i\omega L + \frac{1}{i\omega C}$$

The derivative of the imaginary part of the impedance evaluated at the resonance frequency fixes the value of the inductor and via the resonance frequency also the value of the capacitor. The real part of the impedance at the resonance frequency determines the equivalent resistance. Fig. 6.12 shows the computed impedance of the circuit of Fig. 6.11(b) and the comparison with the equivalent *RLC* circuit. The quality factor of the LC resonator is obtained by considering the equivalent *RLC* circuit's quality factor that is given by

$$Q = \frac{\omega L}{R}$$

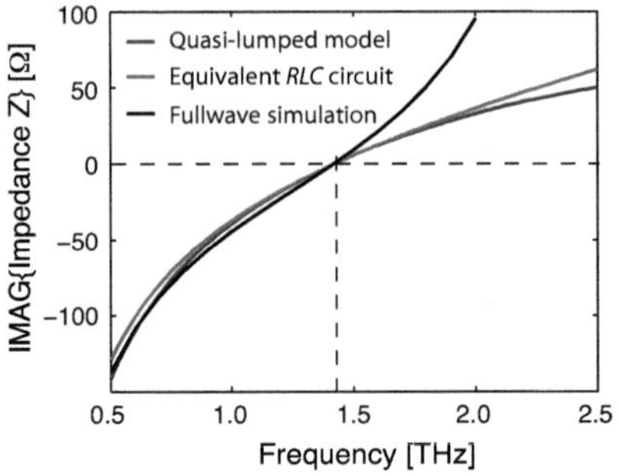

Figure 6.13: Comparison of the imaginary part of the impedance of the microwave model and the equivalent *RLC* circuit with the fullwave simulation. An excellent agreement is observed up to the resonance frequency.

The equivalent *RLC* circuit is given by $R = 1.5\ \Omega$, $L = 5.8$ pH and $C = 2.2$ fF and the computed resonance frequency and quality factor are $f = 1.42$ THz, $Q = 35$. For the geometry defined by Fig. 6.11(a) with the length of the inductor l_1 respectively capacitor l_2 and the width of the inductor w_1 respectively capacitor w_2 following values are used: $l_1 = 10\ \mu$m, $w_1 = 3\ \mu$m, $l_2 = 7.85\ \mu$m, $w_2 = 20\ \mu$m. The complex propagation constant is frequency dependent. At 1.42 THz, it is given for the inductor by $\alpha_1 = 19$ cm^{-1}, $\beta_1 = 8.5 \times 10^2$ cm^{-1}, and the capacitor $\alpha_2 = 7.3$ cm^{-1}, $\beta_2 = 9.1 \times 10^2$ cm^{-1}. The impedance of the inductor section is $Z_{c1} = 64.6\ \Omega$ and the effective dielectric constant is $\varepsilon_{re1} = 8.11$, the impedance of the capacitor section is $Z_{c2} = 25.4\ \Omega$ and the effective dielectric constant is $\varepsilon_{re2} = 9.46$.

6.2.5 Comparison of the 3 models

Three different models for the LC resonator have been discussed. They differ in their complexity, accuracy and computational effort. The first model allows to compute an equivalent LC circuit and to determine the resonance frequency. The simplicity and accuracy of this model is striking but it doesn't allow to compute losses.

The 3D fullwave model gives a complete understanding and all the relevant physical parameters of the resonator. The major drawback being the large computational power, and the demanding memory requirements.

The microwave model allows a detailed analysis of the LC resonator in terms of an equivalent electric circuit. The achieved accuracy for the resonance frequency is usually better than 10% compared to the fullwave simulation, the quality factor differs more. The reason being in the simple approximation for the losses of the complex propagation constant (appendix Sec. B.1). The model is fully analytical, resulting in low computational power and a short calculation time (less than a second). Since it predicts the resonance frequency and the dominant losses it allows for an efficient optimization of design parameters.

Fig. 6.13 shows a comparison of the impedance obtained with fullwave simulation and the microwave model and its equivalent RLC circuit. The very good agreement observed up to the resonance frequency is validating the microwave model in respect to the fullwave simulation. Above the resonance frequency the fullwave simulation and microwave model diverge, this is expected since the fullwave model will find higher order resonances, whereas the microwave model just finds the LC resonance. In principle the microwave model could be adapted to find also the higher order resonances, but it is not of interest in the present context. As a summary the relevant parameters of the LC resonator are given in the table below.

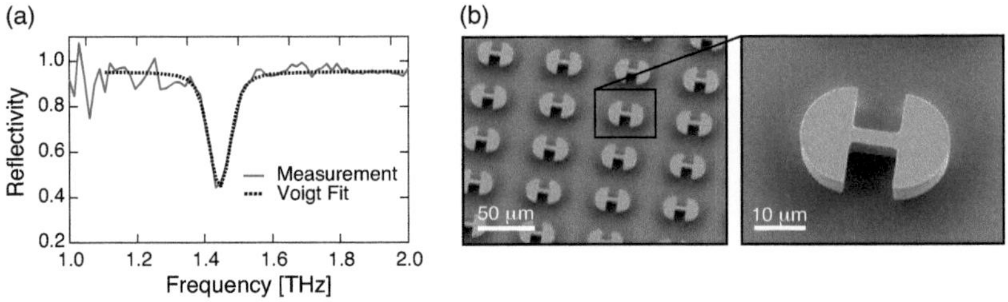

Figure 6.14: (a): Measured reflectivity of an array of 400 identical LC resonators, together with a Voigt fit. (b): SEM picture showing a part of the array and a single LC resonators.

	R [Ω]	L [pH]	C [fF]	f [THz]	Q_{ohm}	Q_{rad}	Q_{tot}	Γ
simple model		4.8	2.3	1.53				
microwave model	1.5	5.8	2.2	1.42	35			
fullwave simulation				1.45	53	189	41	0.85

6.2.6 Measurements

For the validation of the LC resonator concept, the frequency-dependent reflectivity of an array of 400 identical LC resonators with an empty cavity (no active region, undoped GaAs dielectric) has been measured. The setup is described in Sec. 3.2.4. The observed absorption resonance of the reflectivity at 1.45 THz confirms the computed value of the resonance frequency. The linewidth of the absorption would correspond to a quality factor $Q = 20$ at 10 K, which is lower than expected. The lineshape of the absorption is not Lorentzian due to inhomogenous broadening from coupling between the resonators and the fabrication process. Although the Q factor of a single resonator cannot be determined, a Voigt line fit of the absorption line yields a homogenous component corresponding to a quality factor of 40 and an inhomogenous component with a Q of 26. The Q factor of the homogenous broadening would be in good agreement with the computed value for a single LC resonator. The measurement of the quality factor at room temperature (300 K) leads to $Q = 17$ and confirms this interpretation, since the single resonator quality factor is expected to decrease roughly by a factor of 2 between cryogenic temperature and room-temperature due to the higher resistivity of gold [169]. Fig. 6.14 shows a part of the LC resonator array and the measured reflectivity.

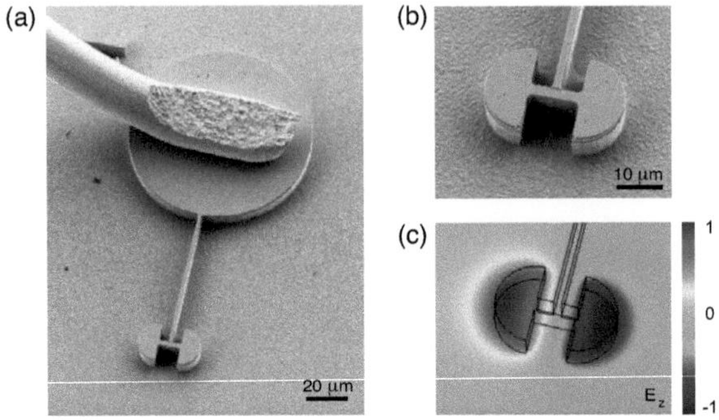

Figure 6.15: (a): Scanning electron micrograph picture of the LC laser including resonator, contact wire and bonding pad. (b): Close up of the resonator. (c): Fullwave simulation including the contact wire showing the electric field E_z. The contact wire is not perturbing the optical mode.

6.3 LC laser

6.3.1 Device fabrication

In contrast to LC resonators, the LC laser is an active device that requires a gain medium and electrical pumping of the latter. A scanning electron micrograph image of the LC laser including a connecting wire to a bonding pad and a close up of the resonator is shown in Fig. 6.15(a,b). An 8μm thick terahertz QCL active region having the gain peaked at 1.5 THz and labeled EV1161 is used. The bandstructure is the same as the one of layer N892 that has been described and studied in detail in chapter 4.4.2. The LC resonator defines naturally a virtual ground for the resonance frequency in the middle of the inductor that is used for the electrical pumping of the active region. Fullwave simulations including a contact wire are performed and show that the optical mode is not influenced by the wire 6.15(c).

Similar fabrication techniques are used for the LC laser than for terahertz QCLs

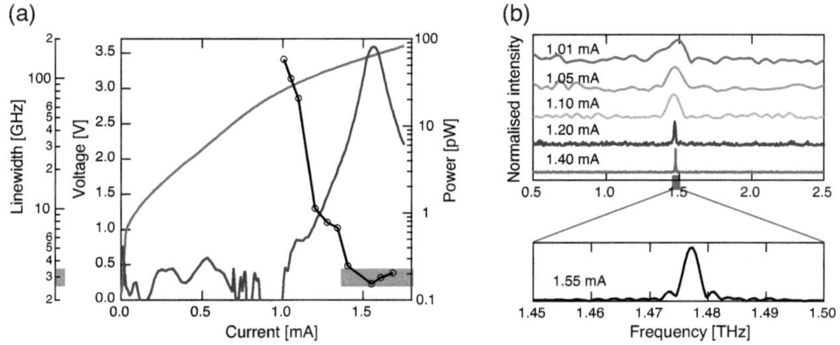

Figure 6.16: (a): Electrical (red curve) and optical (blue curve) characterization at 10 K of the LC laser. The output power is measured with an He-cooled bolometer. Black dots represent the full width at half maximum of the emitted radiation. The emission spectra of the device is measured with a Fourier transform infrared spectrometer, and the linewidth is obtained by a Lorentzian fit of the spectra. The green area indicates the resolution limit of 3 GHz of the spectrometer. (b): Normalized measured spectra at various injection currents corresponding to points in (a).

based on a double-metal waveguide. The resonator and bonding pad are defined by dry etching of the active region with an argon/chlorine based chemistry. The active material below the bonding pad is electrically isolated from the metal to avoid pumping of the latter. The fabrication process is described in detail in Sec. 3.1.3.

6.3.2 Optical characterization

The light and voltage versus current characteristics of a LC laser operated at 10 K show a strong increase in the optical power observed at at current of ~1.20 mA, as the voltage reaches the value corresponding to the alignment of the structure, with a maximum detected peak power of ~80 pW reached at 1.55 mA [Fig. 6.16(a)]. To characterize the nature of the emission, a measurement of the device spectrum as a function of increasing current is performed. The spectra in Fig. 6.16(b) show a single emission line peaking at a frequency of 1.477 THz, close to the value expected by simulation of 1.45 THz. At currents between 1.01 and 1.10 mA, corresponding to the lowest detectable optical signal, the linewidth corresponds to a $Q = 11$ to 21, which is lower than the value computed for the cold cavity. This discrepancy is attributed to two different factors: First, during the alignment regime of the quantum structure near 1 mA, the active region losses may be higher than at the threshold [105]. Second, as the transition occurs between two states with a clear spatial separation, the gain curve exhibits a strong Stark shift with the applied bias. At low biases, the gain peaks at 1.3 THz and is therefore clearly detuned from the cavity resonance, which acts as a filter to further broaden the emission spectrum. The observation of electroluminescence at such low current and frequency points toward a large value of the Purcell factor. As the current is increased, the linewidth decreases steeply and falls below the resolution limit of the Fourier transform infrared spectrometer, corresponding to a linewidth of ~3 GHz at a current of 1.4 mA and a detected power of ~7 pW. As the current is increased beyond this value, any further narrowing of the transition is hidden behind the limitation of the measurement system. the plot of the linewidth versus injected current combined with the super-exponential increase of the optical power in Fig. 6.16 strongly suggests that the device is reaching the laser threshold regime near to 1.5 mA. However, the gain in our structure is not sufficient to operate the device clearly in the regime above the lasing threshold.

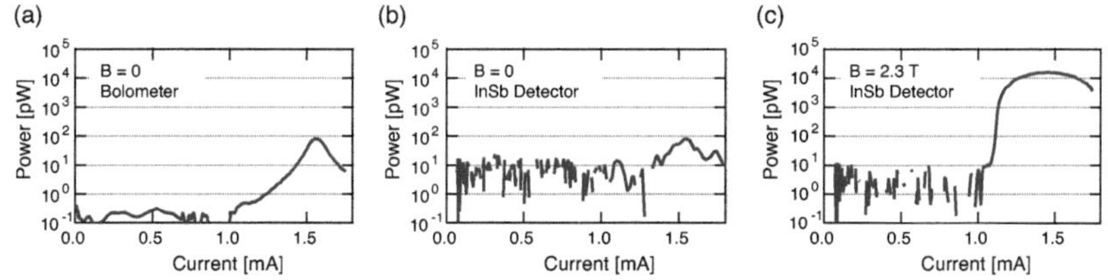

Figure 6.17: Measured power of the LC laser without magnetic field using (a): a bolometer, (b): an InSb detector. (c): Measurement in an applied magnetic field of 2.3 Tesla, and the InSb detector. The noise floor of the bolometer is two orders of magnitude lower than for the InSb detector.

Magnetic field measurements

To access the operation in the regime above threshold, the gain of the active medium is enhanced by immersing the device in a strong magnetic field perpendicular to the quantum wells plane. The magnetic field breaks the in-plane parabolic energy dispersion of the electronic subbands and leads to the formation of discrete Landau levels, allowing the creation of quasi zero-dimensional states. The spacing of the Landau levels is equal to the cyclotron energy $E_c = \hbar eB/m^*$, where $\hbar$ is Planck's constant divided by 2π, B is the magnetic field, e is the electron charge, and m^* is the effective mass of the electron. By varying the relative ratio between the cyclotron energy and subband energy spacing, the electron dynamics and the subsequent lasing properties of a terahertz quantum cascade laser are strongly affected [104]. The optical characteristics of the LC laser in a magnetic field of 2.3 T in Fig. 6.18(a) show that, above the threshold, the power increases linearly with the current up to 16 nW until the electrical rollover is reached. The onset of lasing is accompanied by an abrupt decrease of the differential resistance and is a clear demonstration of the lasing threshold [45]. By measuring the threshold current of

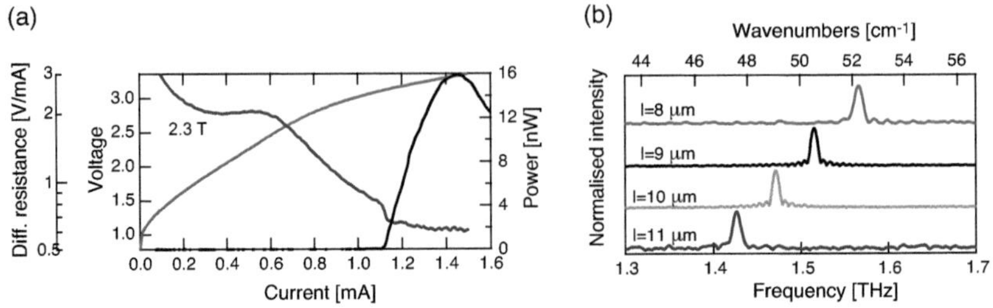

Figure 6.18: (a): Electrical (red) and optical (blue) measurements at 4.2 K for an applied perpendicular magnetic field of 2.3 T. The differential resistance (black curve) shows a large discontinuity at the onset of the lasing threshold at 1.1 mA. (b): Lithographic tuning of the laser frequency by changing the inductor length of the resonator.

a control laser with a conventional metal-metal waveguide, an enhancement of the gain by 5 to 10% is estimated due to the magnetic field of 2.3 T.

Lithographic tuning

A number of resonant cavities have been fabricated, continuously modifying the value of the inductor. As expected, the emission spectra in Fig. 6.18(b) shift correspondingly between 1.43 and 1.57 THz. The spectra of the devices operating in the edge of the gain curve are barely reaching the threshold regime because they are strongly detuned from the maximum of the gain curve.

Current density through the active region

The calculated current density obtained by dividing the measured current through the surface of the LC laser, is to large compared to ridge lasers. The electrical isolation of the bonding pad is not perfect. There is a small, but not negligible leakage current flowing through the area below the pad.

When a metal is making contact with a semicondcutor, a Schottky barrrier is formed

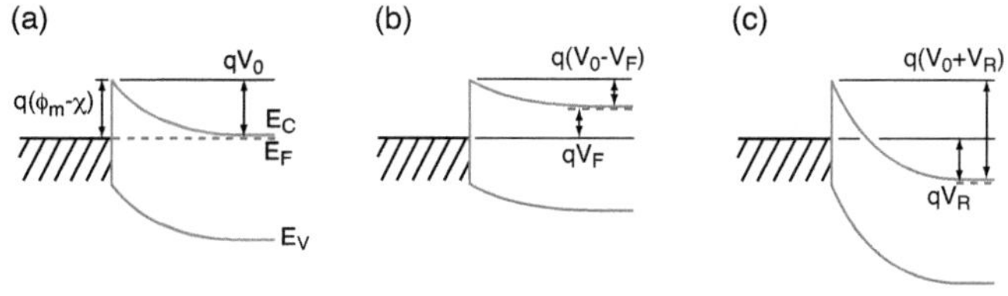

Figure 6.19: Energy-band diagram of a metal n-type semiconductor contact under different biasing conditions. (a): thermal equilibrium, (b): forward bias, (c): reverse bias. ϕ_m is the metal workfunction, χ the electron affinity in the semiconductor, V_F forward and V_R reverse bias.

at the metal-semiconductor interface [170]. At the interface the Fermi level must be continuous across the junction and leads to a transfer of charges between the two materials. Electrons cascade into the metal until the Fermi levels are equalized [Fig. 6.19(a)]. The electron density being much smaller in the semi-conductor than in the metal, an important portion of the semiconductor's volume becomes depleted of electrons when equilibrium is reached forming a depletion layer. A Schottky contact behaves differently depending on the polarization. If a forward bias is applied on a Schottky contact [Fig. 6.19(b)], electrons flow from the semiconductor to the metal. The voltage tends to flatten the barrier seen by the electrons and the Schottky contact becomes conducting at relatively low voltage. If a reverse bias is applied, electrons have to tunnel through the Schottky barrier and the depletion width increases [Fig. 6.19(c)]. The voltage when the junction becomes conductive may be relatively large. For good electrical contacts, the depletion region is minimized by a highly doped layer on the top of the semiconductor leading to a depletion layer in the order of 20 nm for a well designed Au/GaAs contact. The undoped $Al_{0.5}Ga_{0.5}As$ layer below the contact pad leads to a depletion width of more than 200 nm and results in a wide Schottky barrier. The polarity of the active region in the present

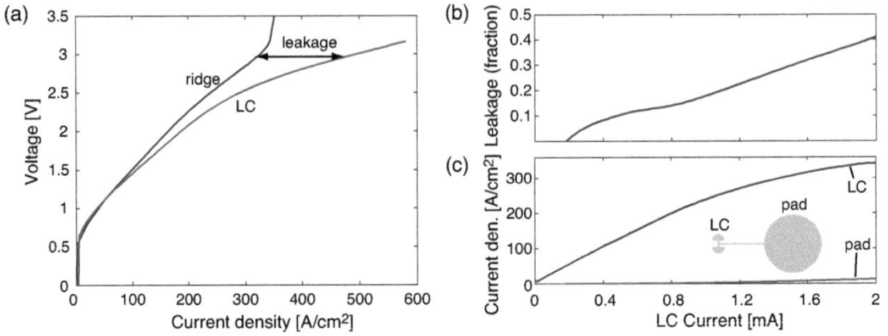

Figure 6.20: (a): Comparison of the electrical characteristics of the LC laser with a ridge laser. The computed apparent current density in the LC laser is larger, due to a leakage current through the contact pad. (b): The leakage current as a fraction of the total current and the (c): deduced current density through the LC laser active region and the contact pad.

LC laser is such that the Schottky contact below the pad is biased in the forward direction. If the polarity of the active region would be inverted, the leakage current could be efficiently reduced.

The leakage current through the pad is deduced from the comparison of the cw measurements of the LC laser and a non lasing ridge laser of the same active layer. (Many process iterations were required for the development of the LC laser. In one of these processes, the ridge lasers have gold sputtered on the sidewalls. The electrical properties are good, but the optical losses are to high for lasing operation.) Fig 6.20(a) shows the comparison of the voltage versus current characteristics. The Schottky voltage at 0 bias varies from one sample to the other. For the comparison, the voltage has been corrected with an offset. The deduced leakage current is shown as a fraction of the total current in Fig. 6.20(b).

Although the fraction of the leakage current is as high as 40%, the current density through the pad is very low since the surface of the pad is 19 times larger than the

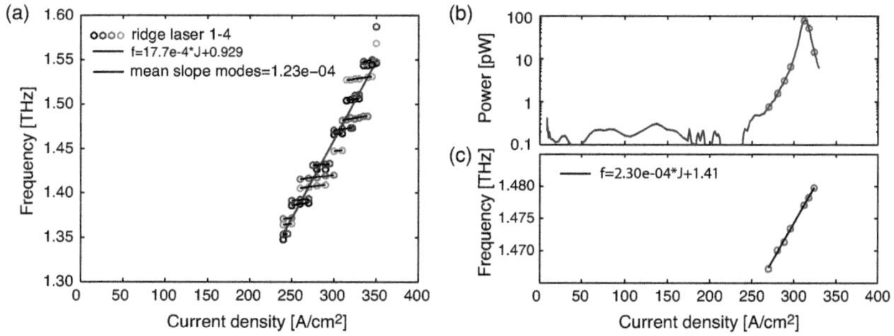

Figure 6.21: (a): Stark-shift of the gain curve, deduced from the emission spectra of 4 ridge lasers, all having dimensions of 1 mm x 100μm. The cavity pulling of the individual laser modes is fitted as well. (b,c): Cavity pulling of the emission frequency of the LC laser, obtained from the measured spectra.

surface of the LC active region. At a current density of $\sim$ 343 A/cm^2 through the LC active region, corresponding to the NDR current density, only a current density of $\sim$ 13 A/cm^2 flows through the pad. Any gain contribution from the active region below the pad can be safely excluded. The real current density through the LC active region and the pad are shown in Fig. 6.20(c). The leakage current is device dependent and has to be considered for every LC laser separately. The data of LC lasers that is shown as a function of the current density is corrected for the leakage, but the data shown as a function of the current is not corrected.

Quality factor Q

The cavity losses of a single mode laser can be determined by exploiting the cavity pulling effect due to the Stark-shift of the gain curve [150, 151, 171]. The cavity pulling effect in lasers is explained in detail in Ref. [43]. A Lorentzian gain results in a dispersion of the refractive index by the Kramers-Kronig relations. In the hot cavity (with gain) the resonance frequency is shifted in respect to the cold cavity (no gain) frequency. It turns out that the lasing frequency is "pulled" from the cold cavity frequency towards the peak gain frequency. The peak material gain g_p is expressed by the formula

$$g_p = \left(\frac{\frac{\partial \nu_{laser}}{\partial J}}{\frac{\partial \nu_{gain}}{\partial J}} \right) \frac{2\pi n \Delta\nu_{gain}}{c} \tag{6.3.7}$$

where $\Delta\nu_{gain}$ is the full width at half maximum of the spontaneous emission and n the refractive index. The spontaneous emission linewidth of the active region used for the LC laser has not been measured. However on a very similar bound-to-continuum active region a value of 1.6 meV at 1.7 THz is measured (Sec. 4.3). For the active region of the LC laser a linewidth of 1.5 meV is assumed, corresponding to a quality factor of 4. The Stark-shift of the gain curve is fitted from the measured emission spectra of several ridge lasers, shown in Fig. 6.21(a). A slope of $\partial\nu/\partial J = 1.77 \times 10^{-3}$ THz/(A/cm^2) is obtained. The shift of the emission frequency of the LC laser by the cavity pulling effect is obtained by measuring high resolution spectra, shown in Fig. 6.21(b,c). The fitted slope is $\partial\nu/\partial J = 2.30 \times 10^{-4}$ THz/(A/cm^2). With this values, a peak material gain of 35.5 cm^{-1} is obtained. It is interesting to note that the Stark-shift and the cavity pulling effect are both linear with the current density. Therefore the peak material gain is constant in the considered range from 270 A/cm^2 to 324 A/cm^2. Since the LC laser works up to the threshold regime, the losses of the LC resonator are estimated to $g_p\Gamma \approx \alpha_{tot}$ (when the gain curve and the LC mode are at resonance). An experimental value for the the quality

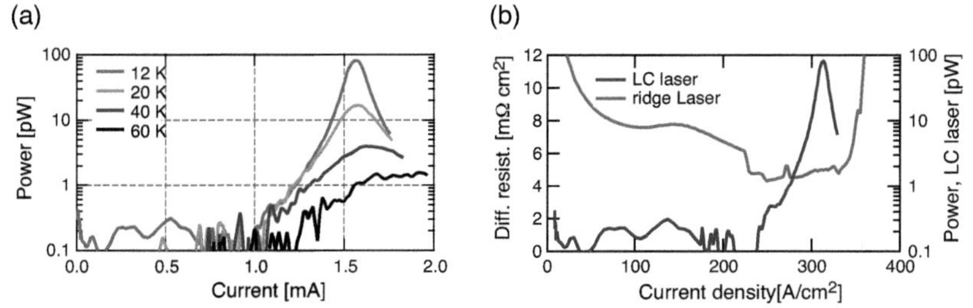

Figure 6.22: (a): Measured power versus current for various temperatures of the LC laser that has been characterized in Fig. 6.16. (b): Comparison of the measured power the LC laser with the differential resistance of a ridge laser with the same active region.

factor of a single LC resonator is then obtained

$$Q_{LC} = \frac{\omega}{\tilde{c}\alpha_{tot}} \tag{6.3.8}$$

With $\Gamma = 0.85$ a quality factor of $Q_{LC} = 37$ is found. This value is close to the theoretical value $Q_{calc} = 41$ obtained when taking into account the computed value for the ohmic and radiative losses $Q_{rad} = 189$ and $Q_{ohm} = 53$. As a by-product the waveguide losses of the ridge lasers are estimated to $\alpha = 19$ cm^{-1} by evaluating the cavity pulling effect on the laser modes in Fig. 6.21(a).

Temperature characterization

The power characteristics of the LC laser has been measured at different temperatures, shown in Fig. 6.22(a). The power of the LC laser shows a strong dependence on the current and temperature since emission below and in the threshold regime is observed. At 40 K and above the measured power corresponds clearly to emission below the threshold regime. For this reason the power dependence of the LC laser on the current and temperature differs from the typical power output of ridge lasers (see data of N892 in Fig. 4.12 on page 94).

It is interesting to note that the onset of spontaneous emission in the LC laser is only slightly above the lasing threshold of a control ridge laser, shown in Fig. 6.22(b). The spontaneous emission can be used as a probe of the upper state population and supports the interpretation in Sec. 4.5.6, that the injection efficiency of carriers in the upper state turns on abruptly, corresponding to the alignment of the injector state with the upper state. The onset of spontaneous emission in the LC laser is almost independent on the current at low temperature. At higher temperature the onset is shifting very quickly to higher currents, similar to what is observed in N892 for the threshold current. This observation suggests that the alignment of the injector with the upper state is pushed to higher currents, leading to higher threshold currents for increased temperatures in the case of the ridge lasers of N892. The higher conductivity prior to the alignment might be the cause of the performance degradation of low frequency terahertz QCLs with temperature.

6.3.3 Emission of circular LC lasers

LC lasers with circular shaped capacitor plates have been fabricated. The frequency is adjusted by changing either the radius r of the capacitor plates or the length l of the inductor as defined in Fig. 6.23(a). A processed device is shown in Fig. 6.23(b). Devices with $r = 6\mu$m and $r = 5\mu$m and various inductor lengths have been characterized and summarized in Fig. 6.24. The maximal measured powers are in the order of $1 - 2$ pW. At the current corresponding to the maximal power emission, narrow spectra are measured. For the $r = 6\mu$m resonators the linewidth

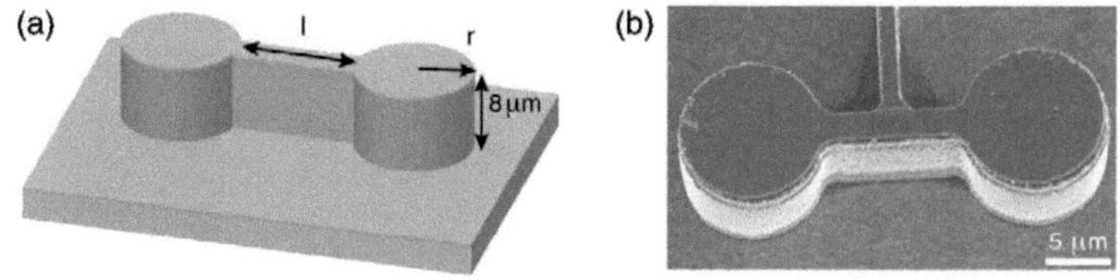

Figure 6.23: (a): Schematic of a LC resonator with circular capacitor plates. (b): A SEM picture of a processed LC laser with circular shaped capacitor plates. The geometry of the resonator is $r = 6\mu$m and $l = 12\mu$m.

is as narrow as $\sim$ 15 GHz, and for the $r = 5\mu$m devices it is in the order of $\sim$ 25 GHz. The threshold regime is not reached in those structures, their linewidths are an order of magnitude wider than for the half-circular shaped LC laser. The emission over a broad current range in Fig. 6.24(a,c) is an artifact of the leakage current through the bonding pad that masks the real current through the active region. The electrical cross section of the active region of the circular LC lasers with $r = 6\mu$m is 75% of the half-circular shaped, respectively 60% for $r = 5\mu$m.

The slightly worse optical properties that are computed for the circular shaped LC lasers, compared to the half-circular shaped, explain the lower measured power and wider linewidth. The ohmic quality factor for the circular shaped LC laser ($r = 6\mu$m, $l = 12\mu$m) determined with the microwave model is $Q_{ohm} = 31$ and the optical confinement factor is $\Gamma = 0.83$. For the half-circular shaped LC laser the computed values are $Q_{ohm} = 35$ and $\Gamma = 0.85$ (See table on page 156).

The measured emission frequency corresponds well to the computed values of the LC resonator frequency. Fig. 6.25 shows an overview of the computed and measured values.

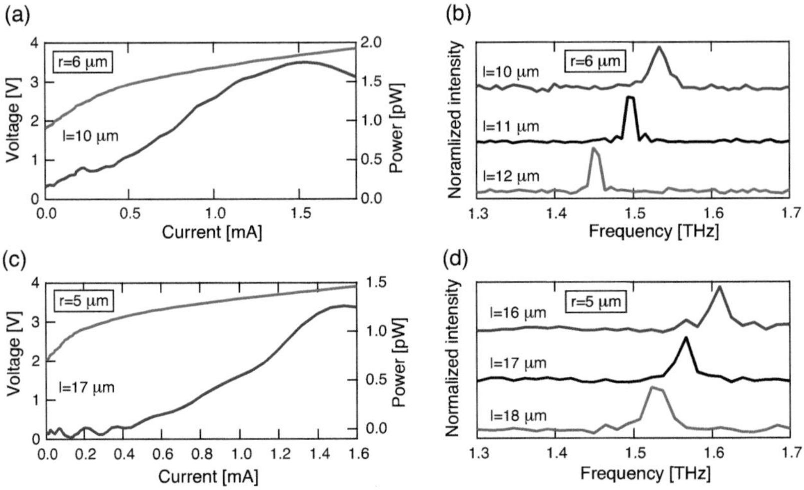

Figure 6.24: (a),(b): Characterization of $r = 6\mu$m LC lasers. The power-current and voltage-current characteristics are shown for a typical device. The spectra measured at the emission maximum are shown for three different lengths of the inductor. (c),(d): Optical and electrical characterization of $r = 5\mu$m devices.

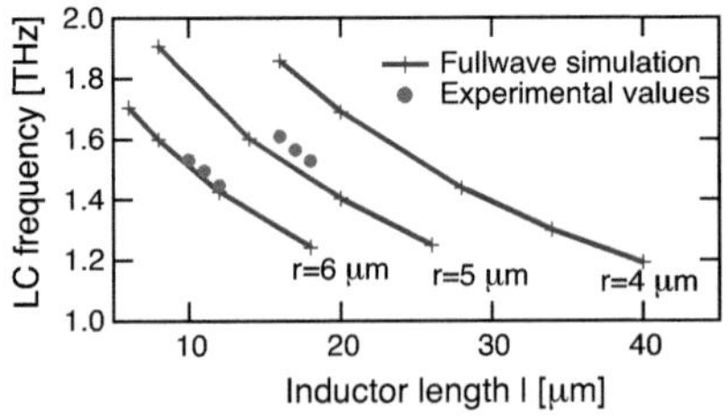

Figure 6.25: Computed LC resonator frequency (fullwave simulation) compared to the measured emission frequency of LC lasers for some geometries.

6.4 Purcell effect in the LC laser

The spontaneous emission rate of a quantum excitation (for example in atoms, quantum wells, dots, etc.) is not an inherent property but arises from the interaction of the quantum system with an infinity of vacuum quantum states. An optical microcavity modifies locally the density of states, and allows to manipulate the dynamics of the spontaneous emission of an excitation in the microcavity [172], in particular to enhance or inhibit the spontaneous emission, the former is also known as the Purcell effect [173], as well as the directionality of the emission. The Purcell effect is observed in semiconductor microcavities with incorporated InAs quantum dots through time resolved photoluminescence measurements, demonstrating a five fold increase of the spontaneous emission rate [174]. Controllable single-photon sources [175], highly desired for applications in quantum information technology, rely on the Purcell effect and high-efficiency microcavity light emitting diodes (LEDs) exploit spontaneous emission angular redistribution [176]. The control of the spontaneous emission through the Purcell effect is a way to reduce the threshold of lasers [177]. The most drastic effect of controlling spontaneous emission in a laser may be the threshold-less laser oscillation in which the light output increases nearly linearly with pump power instead of exhibiting a sharp turn on at a pump threshold [178, 179].

The purpose of this section is to investigate the Purcell effect in the LC laser. The derivation of the Purcell effect is outlined and the theoretical Purcell factor in the LC laser is computed, then the QCL rate equations are expressed in a more convenient form for microcavities. Evidence for a large Purcell factor is obtained through comparison of the experimental data with the predictions from the rate equations.

6.4.1 Light-matter interaction in the weak coupling regime

Interaction Hamiltonian

The dipole interaction hamiltonian of a two level quantum system at position $\mathbf{r}$, weakly coupled to the radiation field of the mode l is expressed as (Ref. [43, 44])

$$\hat{W} = iqE_{max}\left[a_l e^{i\mathbf{k}_l\cdot\mathbf{r}} - a_l^\dagger e^{-i\mathbf{k}_l\cdot\mathbf{r}}\right]\mathbf{f}_l(\mathbf{r})\cdot\mathbf{d} \tag{6.4.9}$$

where q is the charge, $\mathbf{d}$ is the dipole operator acting on the states of the two level system: $\mathbf{d}_{12} = \langle 1|\,\mathbf{d}\,|2\rangle$. The photon creation $a_l^\dagger$ and annihilation a_l operators act on the photon states and are defined by: $a_l\,|n\rangle = \sqrt{n}\,|n-1\rangle$ and $a_l^\dagger\,|n\rangle = \sqrt{n+1}\,|n+1\rangle$, where n is the number of photons in the mode l. $\mathbf{f}(\mathbf{r}) = \mathbf{E}(\mathbf{r})/E_{max}$ is the optical mode spatial function, a vector which describes the local field polarization and relative field amplitude. E_{max} is the maximum field per photon defined by

$$E_{max} = \sqrt{\hbar\omega/(2\varepsilon_0 n_{op}^2 V)} \tag{6.4.10}$$

$$V = \frac{\int n_{op}(\mathbf{r})^2|\mathbf{E}(\mathbf{r})|^2 d^3\mathbf{r}}{\max[n_{op}(\mathbf{r})^2|\mathbf{E}(\mathbf{r})|^2]} \tag{6.4.11}$$

where V is the effective mode volume and n_{op} the refractive index where the maximum field is defined. Eq. 6.4.10 and 6.4.11 can be combined to express the maximum field as

$$E^2_{max} = \frac{\max[n_{op}(\mathbf{r})^2|\mathbf{E}(\mathbf{r})|^2]}{n^2_{op}} \qquad (6.4.12)$$

The radiative transition rate of the quantum excitation from $|2\rangle$ to $|1\rangle$ by emission of a photon into the mode l is obtained by applying Fermis golden rule $\hbar\Gamma_{if} = 2\pi|\hat{W}_{fi}|^2\delta(E_f - E_i)$, where the initial state is $|i\rangle = |2, n\rangle$ and the final state $|f\rangle = |1, n+1\rangle$. Therefore

$$\Gamma^l_{21} = \frac{\pi q^2}{V\hbar\varepsilon}(n+1)\left|\mathbf{f}\cdot\mathbf{d}_{12}\right|^2 \omega\mathcal{L}_e(\omega) \qquad (6.4.13)$$

To take into account the broadening of the emitter the delta function has been replaced by a Lorentzian function $\mathcal{L}_e(\omega)$, centered at $\omega_e = (E_2 - E_1)/\hbar$ and whose surface is normalized to unity. The emission rate is a sum of two terms. The term that is proportional to the photon number n is the stimulated emission, the other term the spontaneous emission.

Purcell effect in a single mode cavity

The spontaneous emission rate of a quantum excitation is computed by summing the emission rates into all allowed photon states. The summation is replaced by an integration over all frequencies, using the density of states. The free space density of states is $\rho_0(\omega) = V\omega^2 n^3_{op}/(\pi^2 c^3)$, and the density of states in a single mode microcavity $\rho_c(\omega)$ is a Lorentzian function, whose surface is normalized to unity, since there is only one mode. The Purcell effect arises from the fact that the density of states in a microcavity can locally exceed the free space density of states

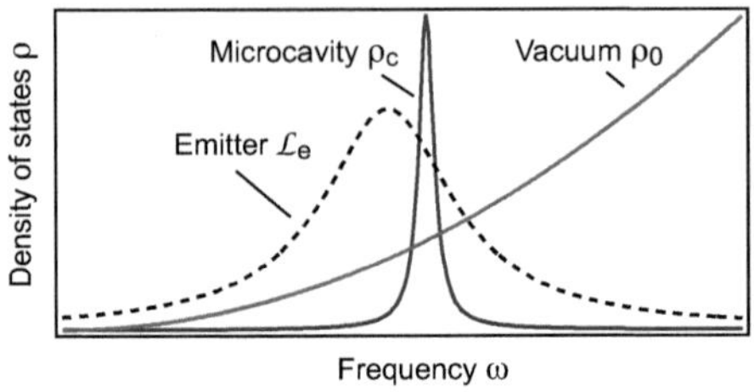

Figure 6.26: The density of states in a single mode microcavity $\rho_c(\omega)$ can be larger than the free space density of states $\rho_0(\omega)$ for certain frequencies. The lineshape of an emitter $\mathcal{L}_e(\omega)$ is shown as well; the density of states of the microcavity and emission lineshape are Lorentzian functions where the width $\Delta\omega$ is related to the respective quality factor $Q = \omega/\Delta\omega$.

for certain frequencies leading to an enhancement of the spontaneous emission rate (see Fig. 6.26). The spontaneous emission term of Eq. 6.4.13 is multiplied by $\rho(\omega)$ and integrated over all frequencies.

$$\Gamma_{sp} = \frac{\pi q^2}{V\hbar\varepsilon}\left\langle|\mathbf{f}\cdot\mathbf{d}_{12}|^2\right\rangle_{modes}\int \omega\rho(\omega)\mathcal{L}_e(\omega)d\omega \qquad (6.4.14)$$

For the free space spontaneous emission rate $\int \omega\rho_0(\omega)\mathcal{L}_e(\omega)d\omega \approx \omega_e\rho_0(\omega_e)$ since $\rho_0(\omega)$ is a slowly varying function compared to $\mathcal{L}_e(\omega)$. The random polarization of free space modes with respect to the dipole results in $\left\langle|\mathbf{f}\cdot\mathbf{d}_{12}|^2\right\rangle_{modes} = d_{12}^2/3$. Hence the free space spontaneous emission rate is

$$\Gamma_0 = \frac{q^2 d_{12}^2\omega_e^3 n_{op}}{3\pi c^3\hbar\varepsilon_0} = \frac{1}{\tau_{sp}^0} \qquad (6.4.15)$$

For the spontaneous emission rate in the cavity the expression is $\int \omega\rho_c(\omega)\mathcal{L}_e(\omega)d\omega \approx \omega_c\mathcal{L}_e(\omega_c)$ since $\mathcal{L}_e(\omega)$ is a slowly varying function compared to $\rho_c(\omega)$ or, equivalently, the linewidth of the cavity is narrower than for the emitter (which is the case in the LC laser, illustrated in Fig. 6.26). Therefore

$$\Gamma_c = \frac{\pi q^2 d_{12}^2}{V\hbar\varepsilon}\xi^2|\mathbf{f}(\mathbf{r}_e)|^2\omega_c\mathcal{L}_e(\omega_c) \tag{6.4.16}$$

where $\xi = |\mathbf{d}_{12} \cdot \mathbf{f}(\mathbf{r}_e)|/(|\mathbf{d}_{12}| \cdot |\mathbf{f}(\mathbf{r}_e)|)$ describes the orientation matching of $\mathbf{d}_{12}$ and $\mathbf{f}(\mathbf{r}_e)$ for the quantum excitation at position $\mathbf{r}_e$. The linewidth function is substituted, $\mathcal{L}_e(\omega_c) = (\Delta\omega_e/2\pi)/[(\omega_c - \omega_e)^2 + (\Delta\omega_e/2)^2]$ where $\Delta\omega_e$ is the full width at half maximum of the emitter linewidth. The ratio of the emission rate in the presence of the cavity by the free space emission rate is obtained in the notation used by Ref. [180]

$$\frac{\Gamma_c}{\Gamma_0} = \frac{3Q_e(\lambda_c/n_{op})^3}{4\pi^2 V}\frac{\Delta\omega_e^2}{4(\omega_c - \omega_e)^2 + \Delta\omega_e^2}\xi^2|\mathbf{f}(\mathbf{r}_e)|^2 \tag{6.4.17}$$

The maximum enhancement of the spontaneous emission is achieved at resonance $\omega_e = \omega_c$, if the emitter is placed in the maxima of the electric field ($|\mathbf{f}(\mathbf{r_e})| = 1$), and the dipole is aligned with the electric field ($\xi = 1$). Then the spontaneous emission rate enhancement is given by

$$\frac{\Gamma_c}{\Gamma_0} = \frac{3Q_e(\lambda_c/n_{op})^3}{4\pi^2 V} = F_p \tag{6.4.18}$$

F_p is called the Purcell factor, from the original derivation of Purcell [173]. Either the cavity mode quality factor Q_c or the emitter quality factor Q_e, whichever is the lower, is limiting F_p. In general a good approximation of the Q in the formula for the Purcell factor in all situations is given when using $1/Q = 1/Q_e + 1/Q_c$.

Purcell effect in a double metal waveguide

An atom placed between two plane mirrors with a sub-wavelength separation remains excited provided that its dipole is oriented parallel to the mirror surfaces [172]. However if the dipole of the atom is oriented perpendicular to the mirror surfaces, its decay rate is larger than in free space [181]. Intersubband excitations in a wide double metal waveguide with a sub-wavelength sized thickness experiences the same modification of the spontaneous emission rate than atoms placed between two plane mirrors. In a wide double metal waveguide, the mode density can be approximated by the mode density of two planar extended mirrors, filled with a dielectricum. If $\lambda/n_{op} > 2L$, where L is the thickness of the active region, the density of states is [181]

$$\rho_{2M} = \frac{V\omega}{2\pi\tilde{c}^2 L} \tag{6.4.19}$$

where $\tilde{c} = c/n_{op}$. The spontaneous emission rate in the double metal waveguide is computed using Eq. 6.4.14, and again $\int \omega\rho_{2M}(\omega)\mathcal{L}_e(\omega)d\omega \approx \omega_e\rho_{2M}(\omega_e)$ since $\rho_{2M}(\omega)$ is a slowly varying function compared to $\mathcal{L}_e(\omega)$. All modes are TM polarized, therefore all the dipoles of the intersubband transition couple to the modes and $\langle|\mathbf{f}\cdot\mathbf{d}_{12}|^2\rangle_{modes} = d_{12}^2$. The Purcell factor in the double metal waveguide is then obtained

$$F_p = \frac{\Gamma_{2M}}{\Gamma_0} = \frac{3\rho_{2M}(\omega)}{\rho_0(\omega)} = \frac{3\lambda}{4n_{op}L} \tag{6.4.20}$$

This formula for the Purcell factor has been used in Ref. [182]. In a double metal waveguide with a thickness of 8 μm the computed Purcell factor is $F_p = 5.2$ at 1.5 THz.

6.4.2 Purcell factor in the LC resonator

The Purcell factor in Eq. 6.4.18 is defined for a single emitter. The active region in the LC laser represents a large statistical ensemble of emitters. To quantify the maximal enhancement of the spontaneous emission at resonance, a statistical average Purcell factor is computed.

$$\bar{F_p} = \frac{3Q_e(\lambda_c/n_{op})^3}{4\pi^2}\frac{\overline{\xi^2|\mathbf{f}(\mathbf{r}_e)|^2}}{V} \tag{6.4.21}$$

Since the dipole of the intersubband transition is polarized along the z direction, the definition of $\mathbf{f}$ and ξ allows to rewrite $\xi^2|\mathbf{f}(\mathbf{r}_e)|^2 = [E_z(\mathbf{r})/E_{max}]^2$. The statistical average is computed by integration over the active region which is essentially the physical volume of the LC resonator.

$$\overline{\xi^2|\mathbf{f}(\mathbf{r}_e)|^2} = \frac{1}{V_{gain}}\frac{\int_{gain} n_{op}(\mathbf{r})^2 E_z(\mathbf{r})^2 d^3\mathbf{r}}{\max[n_{op}(\mathbf{r})^2\mathbf{E}(\mathbf{r})^2]} \tag{6.4.22}$$

The refractive index in the LC resonator is constant and also equal to n_{op} since E_{max} as defined in Eq. 6.4.12 is in the active region. The position dependence of the refractive index is introduced formally for convenience. Eq. 6.4.22 and the effective mode volume V (Eq. 6.4.11) are substituted into Eq. 6.4.21. The average

Purcell factor is then conveniently expressed as

$$\bar{F}_p = \frac{3Q_e(\lambda_c/n_{op})^3}{4\pi^2 V_{gain}}\Gamma \quad (6.4.23)$$

$$\Gamma = \frac{\int_{gain} n_{op}(\mathbf{r})^2 E_z(\mathbf{r})^2 d^3\mathbf{r}}{\int_{all} n_{op}(\mathbf{r})^2 |\mathbf{E}(\mathbf{r})|^2 d^3\mathbf{r}} \quad (6.4.24)$$

Γ is the overlap factor of the optical mode with the gain region defined as the fraction of the electric field energy that couples to the intersubband transition.

A statistically averaged Purcell factor of $\bar{F}_p = 17$ is computed in the LC laser. An emitter linewidth of 1.5 meV is assumed, corresponding to a relatively low $Q_e = 4$. The resonance frequency of the LC resonator is 1.477 THz ($\lambda_c = 203\mu$m), the volume of the LC resonator is $V_{gain} = 2753\ \mu\text{m}^3$ and the overlap factor is $\Gamma = 0.85$. The free space sponatenous emission lifetime of the intersubband transition in the active region is computed with Eq. 6.4.15 and $\tau_{sp}^0 = 28\ \mu$s is obtained ($d_{12} = 10.8$ nm, $f = 1.477$ THz). The spontaneous emission lifetime is expected to get shortened to $\tau_{sp} = 1.6\ \mu$s in the LC resonator, if at resonance, due to the large Purcell factor. However it is still 6 orders of magnitude longer than the non-radiative scattering time of the carriers.

6.4.3 QCL rate equations in a microcavity

To investigate the influence of the Purcell factor on the LC laser properties, it is necessary to reconsider the QCL rate equations. The rate equations 2.2.14-2.2.16 are recalled, but here n_2^{therm} is neglected, and τ_3 is the purely non-radiative lifetime of the upper state that doesn't contain the contribution of the spontaneous emission. τ_{sp} is the spontaneous emission lifetime of the 3 to 2 transition in the presence of the cavity, and the free space spontaneous emission lifetime is noted as τ_{sp}^0 for clarity. The spontaneous emission coupling factor β is defined as the fraction of the spontaneous emission rate into the lasing mode, and $\tilde{c} = c/n_{op}$ is the velocity of light in the material.

$$\frac{dn_3}{dt} = \frac{J}{q} - \frac{n_3}{\tau_3} - \frac{n_3}{\tau_{sp}} - Sg_c(n_3 - n_2) \tag{6.4.25}$$

$$\frac{dn_2}{dt} = \frac{n_3}{\tau_{32}} + \frac{n_3}{\tau_{sp}} + Sg_c(n_3 - n_2) - \frac{n_2}{\tau_2} \tag{6.4.26}$$

$$\frac{dS}{dt} = \tilde{c}\left\{[g_c(n_3 - n_2) - \alpha]S + \beta\frac{n_3}{\tau_{sp}}\right\} \tag{6.4.27}$$

The QCL rate equations are convenient for ridge lasers since the losses α, the gain cross section g_c and the photon flux S are physical quantities that are expressed in length units. However in a microcavity, such as the LC laser, with a non-homogenous cross section, these quantities are ill defined. Instead, the photon lifetime in the cavity τ_c, the stimulated emission rate that depends on the spontaneous emission lifetime τ_{sp}, and the photon number p in the cavity are the natural quantities to describe the laser system.

The photon number p is related to the photon flux per period per stripe width by the definition of the latter $S = \tilde{c}pL/V$, where L is the length of one period of the active region, and the photon lifetime is defined by $\tau_c = 1/(\alpha\tilde{c})$. Using these relations, Eq. 6.4.27 transforms to

$$\frac{dp}{dt} = -\frac{p}{\tau_c} + g_c\tilde{c}L(N_3 - N_2)p + \frac{\beta V N_3}{\tau_{sp}} \tag{6.4.28}$$

N_3 and N_2 are volume densities ($1/\mathrm{m}^3$), related to the sheet densities by $N_{2,3} =$

$n_{2,3}/L$. The volume of the gain medium V_{gain} is noted for simplicity V.

From Einstein's relation between the A and B coefficients it is clear that for every mode, the spontaneous emission equals to the stimulated emission when the average photon number in the mode is unity. The expression $g_c\tilde{c}L$ is identified as $\beta V/\tau_{sp}$, analogue to the derivation in Ref. [183], and the gain cross section is expressed with the spontaneous emission lifetime as

$$g_c = \frac{\beta V}{\tilde{c}L\tau_{sp}} \tag{6.4.29}$$

Finally the QCL rate equations are written with the quantities τ_{sp}, τ_c, and p as

$$\frac{dN_3}{dt} = \frac{J}{qL} - \frac{N_3}{\tau_3} - \frac{p\beta(N_3 - N_2)}{\tau_{sp}} - \frac{N_3}{\tau_{sp}} \tag{6.4.30}$$

$$\frac{dN_2}{dt} = \frac{N_3}{\tau_{32}} - \frac{N_2}{\tau_2} + \frac{p\beta(N_3 - N_2)}{\tau_{sp}} + \frac{N_3}{\tau_{sp}} \tag{6.4.31}$$

$$\frac{dp}{dt} = -\frac{p}{\tau_c} + \frac{p\beta V(N_3 - N_2)}{\tau_{sp}} + \frac{\beta V N_3}{\tau_{sp}} \tag{6.4.32}$$

The gain cross section (Eq. 6.4.29) describes the electron-photon interaction, therefore it shouldn't depend on the cavity. In a large laser cavity, the coupling factor β is inversely proportional to the volume. The spontaneous emission lifetime is independent on the volume and is equal to the free space spontaneous emission lifetime τ_{sp}^0. Therefore the gain cross section is independent on the cavity shape and size. In a small cavity the situation is different. If β is unity and the volume is sufficiently small, then the gain cross section would decrease linearly with the volume. The apparent contradiction is solved by the Purcell effect that is responsible for a decrease of the spontaneous emission lifetime in the cavity. To show this point, the spontaneous emission lifetime in the LC laser is expressed with the average Purcell factor (Eq. 6.4.23), assuming that the LC resonator and the intersubband transition are at resonance.

$$\tau_{sp} = \tau_{sp}^0 / \bar{F}_p = \tau_{sp}^0 \frac{4\pi^2 V}{3Q_e(\lambda_c/n_{op})^3\Gamma} \tag{6.4.33}$$

The spontaneous emission coupling factor is close to unity ($\beta = 1$) in the LC resonator, since there is only one resonant mode within the bandwidth of the emitter and the out-coupling of the resonator is negligible. The spontaneous emission lifetime in the cavity (Eq. 6.4.33), the free space spontaneous emission lifetime (Eq. 6.4.15), and $\beta = 1$, is substituted into the equation for the gain cross section (Eq. 6.4.29). The gain cross section as defined in Eq. 2.2.11 on page 27 is recovered.

$$g_c = \frac{\Gamma 4\pi q^2 d_{12}^2}{\varepsilon_0 n_{op} \lambda 2\gamma_{12} L_p} \tag{6.4.34}$$

The fact that the threshold condition is reached in the LC laser, is an indirect proof for the Purcell effect and the shortening of the spontaneous emission lifetime in the LC laser. Since the mode volume is strongly sub-wavelength, the gain cross section would be about 17 times smaller than in free space, or more than 3 times smaller than for ridge waveguides if there was no Purcell effect. To investigate the Purcell effect on the light output characteristics of the LC laser, the rate equations must be solved and compared to the experimental data.

6.4.4 Comparison with experiment

Solution of the rate equations

The QCL rate equations 6.4.30-6.4.32 are solved in the steady state regime for the LC laser ($\beta = 1$). The detailed calculations are performed in the appendix Sec. C.1. The photon number as a function of the inversion ratio r, with the definitions of the saturation photon number p_{sat} and the factor f, writes as

$$p = \left((r-1) + \sqrt{(r-1)^2 + \frac{4rf}{p_{sat}}} \right) \frac{p_{sat}}{2} \tag{6.4.35}$$

$$\begin{aligned} \text{where} \qquad r &= J/J_{th} && (6.4.36) \\ p_{sat} &= \tau_{sp}/(\tau_2 + \tau_{eff}) && (6.4.37) \\ f &= \tau_3/\tau_{eff} && (6.4.38) \\ J_{th} &= \frac{qL}{V\tau_c}\frac{\tau_{sp}}{\tau_{eff}} && (6.4.39) \\ \tau_{eff} &= \tau_3(1 - \tau_2/\tau_{32}) && (6.4.40) \end{aligned}$$

The threshold condition is reached for $r = 1$. The photon number function $p(r)$ is characterized by a strong step-like increase from the spontaneous emission dominated regime ($r < 1$) to the lasing regime, dominated by stimulated emission ($r > 1$). The photon number function is well approximated by following expressions, as long as $r = 1 \pm \delta$ is not too close to 1, e.g. $4rf/p_{sat} \ll \delta^2$, being satisfied in the LC laser if $0.01 \leq \delta$. (Appendix Sec. C.2)

$$\text{if } r < 0.99: \qquad p \approx \frac{r}{1-r} f \tag{6.4.41}$$

$$\text{if } r > 1.01: \qquad p \approx (r-1) p_{sat} \tag{6.4.42}$$

The efficiency of generating photons below threshold is governed by f, and above threshold by p_{sat}. The ratio p_{sat}/f is a measure for the step-like increase of the photon number at the threshold.

$$\frac{p_{sat}}{f} = \frac{\tau_{sp}}{\tau_3} \frac{\tau_{eff}}{\tau_2 + \tau_{eff}} \tag{6.4.43}$$

The expression is proportional to the spontaneous emission lifetime and therefore inversely proportion to the Purcell factor. The dependence on τ_{eff} and therefore on τ_2 and τ_{32} comes from the non unitary internal quantum efficiency in the lasing regime, that has been identified in Eq. 2.2.19 as $\tau_{eff}/(\tau_2 + \tau_{eff})$.

Purcell factor extracted from the fit of the threshold region

The measured power as a function of the current density of the LC laser in the magnetic field is fitted by the photon number function to get the best fit in the threshold region, using p_{sat}/f as a fit parameters. The measured power P is is related to the photon number p by

$$P = h\nu \frac{p}{\tau_{rad}} \eta_{opt} \tag{6.4.44}$$

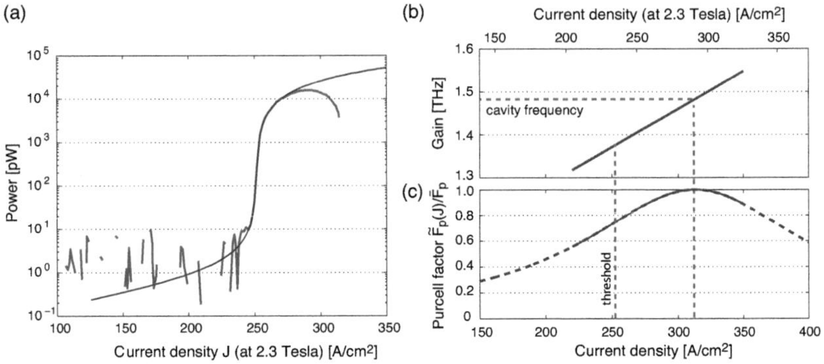

Figure 6.27: (a): Fit of the measured power of the LC laser in the magnetic field ($B = 2.3$ T) with Eq. 6.4.35. The fit parameter is independent on the optical collection efficiency. (b): The Stark-shift of the gain curve as a function of the current density and (c): the Purcell enhancement $\tilde{F}_p$ normalized by the maximal value at resonance $\bar{F}_p$ as a function of the current density.

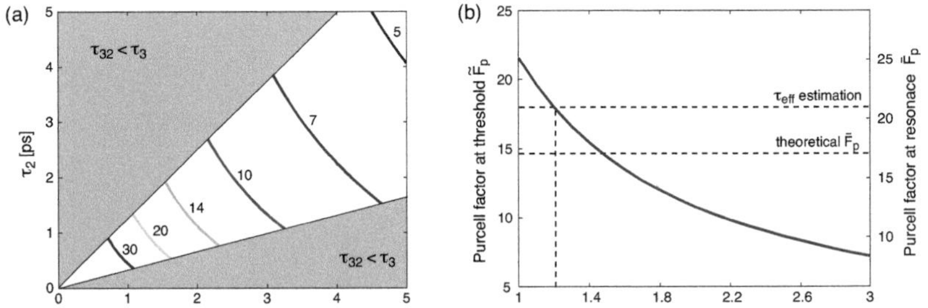

Figure 6.28: (a): The deduced Purcell factor at threshold $\tilde{F}_p$ as a function of the lifetimes τ_3 and τ_2. The grey area is unphysical since in this area $\tau_{32} < \tau_3$. (b): The deduced Purcell factor as a function of the effective lifetime τ_{eff}.

where η_{opt} is the optical collection efficiency. Multiplying the photon number p in Eq. 6.4.35 by a factor η is equivalent to multiply p_{sat} and f by the same factor η. p_{sat} and f cannot be determined independently, since the optical collection efficiency is unknown, but their ratio p_{sat}/f can be deduced from the fit. Fig. 6.27(a) shows the measured power versus current density data together with the fit. An excellent fit is obtained when the data exceeds the noise floor, in particular in the threshold region and above threshold. Two experimentally determined coefficients are defined for convenience

$$\varphi = \frac{p_{sat}}{f} \tag{6.4.45}$$

$$\chi = \frac{J_{th} V \tau_c}{qL} \tag{6.4.46}$$

φ is the fit parameter and χ is determined through the measurement of the threshold current density. τ_c is determined with the experimental value of the quality factor $Q = 37$. Following values are obtained: $\varphi = 5.68 \times 10^5$ and $\chi = 1.28 \times 10^6$.

The fitted curve in Fig. 6.27(a) doesn't correspond to a unique set of values for the lifetimes τ_3, τ_2, τ_{32}, τ_{sp}. The Purcell factor at the threshold $\tilde{F}_p$ is deduced from the subset of lifetimes that reproduces the fitted curve. Mathematically speaking, there are four unknown lifetimes and two equations through the measurement of p_{sat}/f and J_{th} (Eq. 6.4.37-6.4.39). The Purcell factor as a function of τ_3 and τ_2 is (appendix Sec. C.3)

$$\tilde{F}_p(\tau_3, \tau_2) = \frac{2\tau_{sp}^0}{\varphi \tau_3} \left[1 + \sqrt{1 + 4 \frac{\chi}{\varphi} \frac{\tau_2}{\tau_3}} \right]^{-1} \tag{6.4.47}$$

(6.4.48)

Fig. 6.28(a) shows the Plot of the Purcell factor $\tilde{F}_p$ as a function of τ_3 and τ_2. The Purcell factor can also be expressed as a function of τ_{eff} (Eq. 6.4.39 and 6.4.46)

$$\tilde{F}_p(\tau_{eff}) = \frac{\tau_{sp}^0}{\chi\tau_{eff}} \tag{6.4.49}$$

The plot of the Purcell factor as a function of τ_{eff} is shown in Fig. 6.28(b). The effective lifetime τ_{eff} in the LC active region is estimated through the measurement of the peak material gain with the cavity pulling effect

$$\tau_{eff} = \frac{qg_p}{Jg_c} \tag{6.4.50}$$

where g_p is the peak material gain and g_c the calculated gain cross section at resonance. Following values have been used: $g_p = 35.5$ cm^{-1}, $g_c = 1.76 \times 10^{-8}$ cm, $J = 270$ A/cm^2. The deduced $\tau_{eff} = 1.2$ ps represents a lower limit since the lifetimes are larger in the magnetic field at 2.3 Tesla. The theoretical value for the Purcell factor $\bar{F}_p = 17$ agrees with the deduced upper limit $\bar{F}_p = 21$.

Note that the Purcell factor at the threshold $\tilde{F}_p$ is not the Purcell factor of the definition 6.4.23, since the latter corresponds to the maximum enhancement of the spontaneous emission when the emitter is in resonance with the cavity. This is not the case at threshold in the present LC laser, due to the strong Stark-shift of the gain curve with the voltage, respectively current. The dependence of the emission frequency on the current density has been established through the study of the emission spectra of ridge lasers in Sec. 6.3.2. The gain frequency as a function of the current density is shown in Fig. 6.27(b) and the corresponding Purcell factor

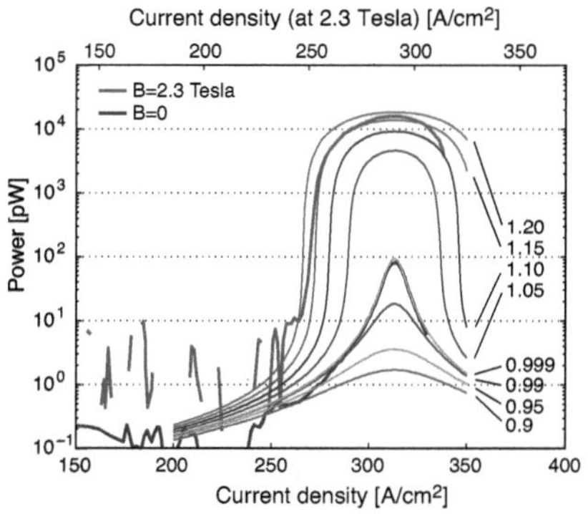

Figure 6.29: Measured power versus current density curves of the LC laser in the magnetic field (2.3 T) and without magnetic field, together with the ploted power for different maximal inversion ratios r_{max}. The inversion ratio is controlled by the J dependence of the Purcell factor $\tilde{F}_p(J)$. Following values have been assumed for the lifetimes: $\tau_3 = 2$ ps, τ_2 =1 ps, and therefore $\bar{F}_p = 17$.

$\tilde{F}_p$ rationed by its maximal value $\bar{F}_p$ is shown in Fig. 6.27(c). The extrapolated frequency at the threshold of the LC laser is 1.408 THz. The enhancement of the spontaneous emission at threshold is therefore $\tilde{F}_p \sim 0.858\bar{F}_p$.

Region above the threshold

An important observation is that the fit in Fig. 6.27(a) differs significantly from the measured data well above the threshold. The power drops before the NDR region is reached and it has almost a symmetric shape in respect to the maximum emission. In Sec. 6.3.2 a constant peak material gain has been deduced between 270 A/cm^2 and 324 A/cm^2 in the active region of the LC laser. This points out that the population inversion doesn't increase with the current density in this region. Either the injection efficiency is not unity throughout this region or the lifetimes change as a function of the carrier density. Although the peak material gain is constant, the gain cross section of the LC mode is dependent on the current density through the Purcell factor $\tilde{F}_p(J)$ by Eq. 6.4.29. Making use of Eq. 6.4.36, 6.4.39, and 6.4.50 the inversion ratio writes

$$r = \frac{g_p V \tau_c}{g_c L \tau_{sp}^0} \tilde{F}_p(J) \propto \tilde{F}_p(J) \tag{6.4.51}$$

r is proportional to $\tilde{F}_p(J)$ since the peak gain g_p is constant. (g_c is the gain cross section at resonance.) The photon number is computed and compared to the measured power of the LC laser with and without magnetic field. Different values for the maximal inversion ration are assumed. The magnetic field results in longer lifetimes, therefore a larger inversion ratio is reached than without magnetic field. In Fig. 6.29(c) a good qualitative agreement is found between the measured power characteristics in the magnetic field, when the maximal inversion is $r_{max} \sim 1.15$, and without magnetic field when $r_{max} \sim 0.999$, therefore very close to the threshold. Note that the the computed curves differ qualitatively in the low current density regime and in the threshold region. In these areas the inversion ratio is not only controlled by $\tilde{F}_p(J)$ since the peak material gain also depends on the current density.

6.5 Outlook

The LC laser is only one example of a whole class of applications opened by the LC resonator. The very high confinement of the electric field it enables makes it extremely attractive for detectors, modulators, and also for quantum optic studies. A few ideas for future research projects based on the LC resonator concept are briefly described in this section. An international patent application (PCT) has also been submitted for some of the possible applications presented here.

6.5.1 Engineering of the LC resonator

Scaling of the mode volume

The mode volume of the LC resonator is about $0.12(\lambda/2n)^3$ and the resonator exhibits a total quality factor of 41. Since the LC resonator consists of lumped elements, in principle, its volume should be scalable almost arbitrarily. A straight forward reduction of the lateral size and the active region thickness down to 2 μm allows indeed the scaling of the mode volume to 1/6 of the original LC resonator, shown in Fig. 6.30(a,b). However the quality factor of the scaled resonator is reduced to 26 ($Q_{ohm} = 27$, $Q_{rad} = 1300$), limited by the ohmic losses, and the confinement factor is $\Gamma = 0.93$. A further scaling of the LC resonator's mode volume is possible, but on the cost of an even lower quality factor.

A much more favorable scaling is achieved if the inductor is realized as a bent wire. Fig. 6.30(c) shows a 3D LC resonator using a bent wire and a 0.8μm thick active region. The mode volume is only $0.002(\lambda/2n)^3$, therefore 60 times smaller than the mode volume of the actual LC resonator. The total quality factor is as large

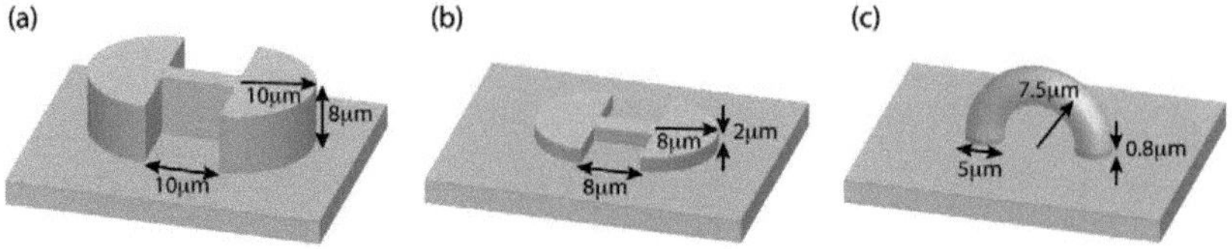

Figure 6.30: (a): Resonator of the LC laser, the reduced mode volume is $V_r = 0.12$, the quality factor $Q = 41$. (b): Scaled resonator, $V_r = 0.02$, $Q = 26$. (c): Resonator with a 3D wire, $V_r = 0.002$, $Q = 63$.

as 63 ($Q_{ohm} = 97$, $Q_{rad} = 183$) and the confinement factor is $\Gamma = 0.85$. The wire corresponds, if unrolled, to an inductor that has a length of 24 μm and a width of 16 μm. The cross section of the wire seen by the surface currents is significantly larger for a 3D wire than for a planar wire, leading to lower ohmic losses. The 3D LC resonator is interesting for lower ohmic losses, smaller mode volumes but also for the low radiative Q.

Radiative Q

An antenna's electrical size is defined in terms of its occupied volume relative to the operating wavelength. An electrically small antenna is one whose overall occupied volume is such that ka is less than or equal to 0.5, where k is the free-space wavenumber $2\pi/\lambda_0$, and a is the radius of a sphere circumscribing the maximum dimension of the antenna [184]. If the antenna is located on a ground plane structure, the antenna image must be included in the definition of a. The LC resonator acts as an electrically small antenna.

There is a theoretical limit for the minimum radiative Q of electrically small antennas. The Fundamental limit on the performance of small antennas was first addressed by Wheeler in 1947 [185]. Wheeler preferred to characterize fundamental limits on the performance of the small antenna in terms of its radiation power factor, which he defined as the inverse of Q. The minimum achievable Q, or lower bound is

$$Q_{Wheeler} = \frac{1}{(ka)^3} \tag{6.5.52}$$

Subsequent work by Chu [186] and Mc Lean [187] has led to the commonly accepted lower bound on Q (the Chu limit) being defined as

$$Q_{Chu} = \frac{1}{(ka)^3} + \frac{1}{ka} \tag{6.5.53}$$

Wheeler's and Chu's formula state the lower bound on Q for the lossless antenna. At small values of ka, the difference between the Chu and Wheeler limit is not significant.

Fig. 6.31 shows the computed Chu limit for small antennas as a function of the electrical size and compared to the radiative quality factor of different planar LC resonators [Fig. 6.30(a)] and 3D LC resonators [Fig. 6.30(c)]. For the computation of the radiative Q of the two types of LC resonators, the height of the capacitors is used as a parameter; frequency and electrical size depend on the height, whereas a is almost constant. As expected, the radiative Q of the LC resonators decreases with increasing electrical size. The planar LC resonators have a large radiative Q , $1-2$ orders of magnitude larger than the Chu limit. The 3D LC resonators have a lower radiative Q, about $0.5-1$ order of magnitude larger than the Chu limit. In conclusion, the LC resonators are not optimized for minimal radiative Q. However

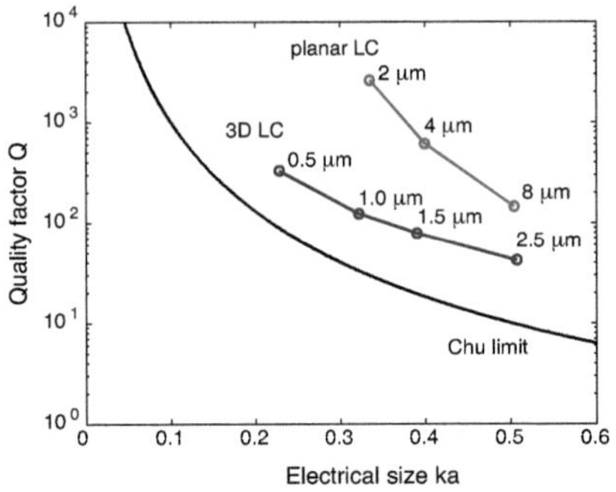

Figure 6.31: The computed radiative Q of planar LC resonators [Fig. 6.30(a)] and 3D LC resonators [Fig. 6.30(c)] as a function of the active region height. The variation in height results in a change in frequency and therefore in ka. The Chu limit gives a fundamental limit for the lower bound of the radiative Q.

the radiative Q can be tuned by the electrical size of the resonator, and in particular 3D LC resonators show a lower radiative Q than planar LC's, while at the same time the ohmic Q may be larger, leading to a higher radiation efficiency in 3D LC resonators.

Far-field

The far-field of the LC laser is computed using the built-in function of Comsol, available in the harmonic propagation analysis. A comparison of the far-field with that of a magnetic half-loop antenna (r = 7.6μm) on a ground plane shows a good qualitative agreement, shown in Fig. 6.32. The far-field of the magnetic half-loop antenna is computed using the fomulas from antenna theory [184]. Obviously, the sub-wavelength nature of the LC laser results in a wide spread far-field. The comparison suggests that the radiation mechanism even of a planar LC resonator is similar to a magnetic half-loop antenna.

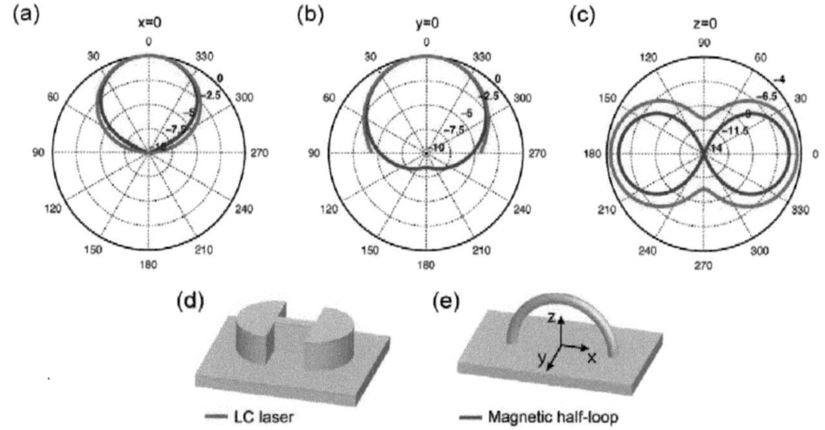

Figure 6.32: (a)-(c): The radiation pattern of the LC laser (d) compared to a magnetic half-loop antenna (e). The far-field is plotted in logarithmic units on three cuts, corresponding to the usual representation in antenna theory.

Frequency scaling

The concept of the LC resonator can also be applied at shorter wavelengths, at mid-infrared or even telecom wavelengths. Fig. 6.33(a) shows two LC resonator designed for 200 THz ($\lambda = 1.5\ \mu$m). For the simulation, a semiconductor refractive index of 3.6 is used and the refractive index of gold at 200 THz is $n = 0.384 + i10.9$ [188]. The planar LC resonator has a radiative quality factor of $Q_{rad} = 15.5$ and an ohmic quality factor of $Q_{ohm} = 28.8$ resulting in a total quality factor of $Q_{tot} = 10.4$. The mode volume is $V_{eff} = 0.14(\lambda/(2n))^3$. The 3D LC resonator has a radiative quality factor of $Q_{rad} = 34.6$ and an ohmic quality factor of $Q_{ohm} = 21.7$ resulting in a total quality factor of $Q_{tot} = 13.5$. The mode volume is $V_{eff} = 0.026(\lambda/(2n))^3$. The lower radiative quality factor of the planar LC resonator is due to the larger electrical size of $ka = 0.76$ compared to $ka = 0.47$ for the 3D LC resonator.

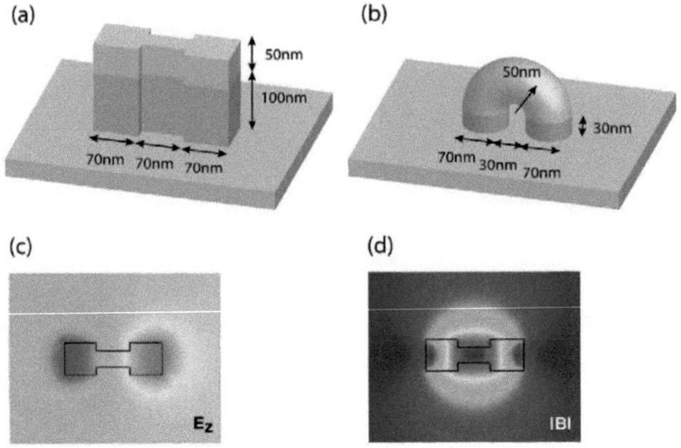

Figure 6.33: Geometry of LC resonators designed for $\lambda = 1.5\ \mu$m in (a): planar and (b): 3D configuration. (c) E_z component of the electric field and (d) the norm of the magnetic field of the planar LC resonator on a plane across the semiconductor material at half substrate height.

6.5.2 Lasers

Local oscillator

Local oscillators (LO) for terahertz heterodyne receivers need very view power [189, 190]. Typically LO powers in the order of 1 μW are sufficient to pump hot-electron bolometer mixers. In a heterodyne receiver, the LO radiation is coupled through an optical beam path and an antenna to the hot-electron mixer resulting in a low coupling efficiency. An LC laser could be used as a local oscillator, due to its low power consumption it could be integrated monolithically with the mixer. The current in the inductor could be coupled directly into the mixer. The resulting device would have the advantage of small size and could be integrated into large arrays.

Multicolor sources

Sub-wavelength multicolor sources can be fabricated by integration of several LC lasers with different resonator frequency. Electrical switching between the different LC lasers would allow to change the emission wavelength - but the optical system wouldn't require a realignment, since the source is sub-wavelength.

Frequency selective components

A combination of optical and microwave components could lead to devices with new functionalities. Almost any kind of filters can be fabricated with microstrip-line components [117]. Filters in combination with double metal waveguides could be used for frequency selection in single mode lasers. Coupling of filters to double-metal waveguides could be achieved with Klopfenstein tapers [191]. A Klopfenstein taper is a continuous impedance transformer between two microstrip-lines of different width, therefore different impedance. It has negligible return loss. A narrow-band anti-reflection coating for double-metal waveguides of the fundamental TEM mode could be realized using a Klopfenstein taper to match a double metal waveguide to a patch antenna. Tunability of filters could be achieved similar to active metamaterials by making use of Schottky depletion [192]. Maybe a combination of such microwave circuits and double metal waveguide terahertz QCL's could lead to voltage tunable terahertz QCL's, or at least to single mode emission with voltage controlled mode-hope tuning.

6.5.3 Strong coupling

The interaction of light with material excitations can be completely altered in microcavity structures where the electromagnetic field is strongly confined and the photonic density of states deeply modified. In particular, if the coupling between the optical transition and the radiation is sufficiently strong compared to the damping rates, new elementary excitations, eigenstates of the full photon-matter Hamiltonian, take form, usually named: cavity polaritons. They present a characteristic anticrossing behavior, with a mode separation often referred to as vacuum-field Rabi splitting, in analogy to the well-known Rabi splitting of saturable two-level systems. The study of these phenomena started some time ago in atomic physics [172]. In the semiconductor world, cavity electrodynamics has developed mainly in conjunction with excitonic states, albeit in a variety of systems, from quantum wells [193, 194, 195, 196] to bulk [197], from III-V compounds to II-VI [198]. Intersubband polaritons, the strong coupling of light with an intersubband transition, has been observed in the mid-infrared wavelength range [199, 200] using dielectric waveguides based on total internal reflection. In the terahertz range the strong coupling is observed using a double-metal waveguide configuration for the light confinement [201]. The Rabi frequency Ω that gives the strength of the light-matter coupling is given by (Ref. [180])

$$\hbar\Omega = |E_{max}\,\mathbf{d}\cdot\mathbf{f}(\mathbf{r}_e)|$$

where the material excitation is at the location $\mathbf{r}_e$ with an electric dipole $\mathbf{d}$. $\mathbf{f}(\mathbf{r}) = \mathbf{E}(\mathbf{r})/E_{max}$ is the optical mode spatial function which describes the local field polar-

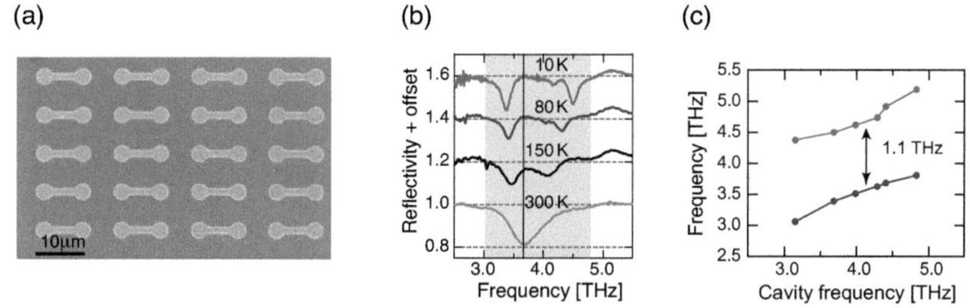

Figure 6.34: (a): Array of LC resonators filling a 1mm x 1mm square. (b): Observation of intersubband polaritons below 150 K in reflection measurement. (c): Tracing of the upper and lower polariton branch. A Rabi frequency of 1.1 THz is extracted.

ization and relative field amplitude. E_{max} is the maximum field per photon already defined in Eq. 6.4.10.

The LC resonator is a nice cavity to study terahertz intersubband polaritons due to its high confinement factor, small mode volume and its quasi single mode spectrum. A series of modulation doped quantum wells is used as the active medium in the LC resonator for the observation of intersubband polaritons in the reflection mode. The sample active region consists of 15 quantum wells, each modulation doped with a sheet density of $n_s = 1 \times 10^{11}$ cm^{-3}. The intersubband transition is designed for 4.1 THz, including the depolarization shift. Large arrays of LC resonators with a circular shape for the capacitor are fabricated, shown in Fig. 6.34(a). The reflection of a broadband terahertz source on the LC resonator array is measured. At 10 K two well pronounced dips are visible in the reflection spectrum, corresponding to the upper and lower polariton, Fig. 6.34(b). At higher temperature, the splitting between the two polaritons is lower, since the electrons are thermally redistributed in the quantum wells, resulting in a lower coupling. Polaritons are observed up to about 150 K. At room temperature the reflection spectrum shows only one dip due to the LC resonator mode. The LC resonator frequency is tuned lithographically. Measuring arrays of resonators at different frequencies, the upper and lower polariton branch is sampled, shown in Fig. 6.34(c). From the minimal splitting between the polariton branches a Rabi frequency of about 1.1 THz is extracted. Improvement on the active region has led to the observation of intersubband polaritons at room temperature using a parabolic well active region in a LC resonator [202].

6.5.4 Detectors

Quantum well infrared photodetectors (QWIPs) are based on intersubband absorption in quantum wells. They have good performances at a variety of different wavelengths [203]. Two major QWIP schemes exist. Photovoltaic QWIPs generate a bias when illuminated, while photoconductive QWIPs are biased with an external voltage and photoemission of electrons from quantum wells leads to a photocurrent. QWIPs have been demonstrated in the terahertz range [204].

An important aspect of QWIPs is the coupling of the incident radiation to the QWIP active region. The intersubband transition selection rule requires a nonzero polarization component in the quantum well direction (epitaxial growth direction). Therefore radiation in normal incidence on the sample surface doesn't couple to the intersubband transition, requiring other coupling geometries. A common coupling scheme is through polished facets at an angle of 45 degrees, forming a single or multipass zigzag waveguide.

For the terahertz region, the coupling of radiation to the active region of photodetectors is very inefficient through the coupling schemes described above due to the long wavelength (typically 100 μm) and the thin active region (typically $5-10$ μm).

The dimension mismatch results in a short effective interaction length.

A more efficient coupling in the terahertz range could be achieved using an LC resonator with an active region of a QWIP. The LC resonator consists ideally of a 3D wire, that has a low radiative quality factor Q_{rad}. In such a configuration the incident radiation couples at normal incidence, a preferred geometry. A stronger interaction of light with the detector active region is provided, since the optical mode of the resonator has a high confinement factor (80-90%). The intensity of the incident power that is coupled to the active region is higher. This, and the narrow bandwidth of the LC resonator leads to a larger signal to noise ratio. A LC based QWIP detector is smaller than the wavelength allowing for integration into dense arrays in contrast to detectors based on the common coupling schemes described above.

6.5.5 Terahertz emitter

A terahertz emitter generates narrowband, but incoherent light. For many applications this may be sufficient. The generation of incoherent narrowband radiation may be feasible at room temperature using spontaneous emission of radiation of an intersubband transition in a quantum well enhanced by a large Purcell effect. The active region of a Purcell emitter is based on a quantum well system with an intersubband transition where electrons are electrically injected in the higher energy level and extracted from the lower energy level. The emitter is based on the radiative transition from the upper to the lower energy level characterized as spontaneous emission. A population inversion is favorable. In principle, it operates also with equal populations in the upper and lower state. The active region is then transparent for emitted photons. The only issue could be Pauli blocking at higher electron densities.

Spontaneous emission is a very slow process, the lifetimes being in the μs range for terahertz frequencies. The non-radative transitions have lifetimes in the order of a few ps, making the spontaneous emission an inherently inefficient process. The spontaneous emission process can be modified due to the Purcell effect when the emitter active region is placed in a resonator that has a very small volume for the optical mode and a large quality factor. In such a situation, the spontaneous emission process is speeded up, and hence the efficiency of photon generation is increased.

The conversion efficiency of electrical power into narrowband terahertz radiation can be described by following equations. A unit pumping efficiency of carriers in the upper state is assumed. Recalling the average Purcell factor in the LC resonator (Eq. 6.4.23), the internal quantum efficiency is

$$\eta_{int} = \frac{1}{1+\tau_{sp}^0/(\tau_{nr}F_p)} \approx \frac{\tau_{nr}}{\tau_{sp}^0}\bar{F}_p \tag{6.5.54}$$

$$\bar{F}_p = \frac{6}{\pi^2}\frac{(\lambda/2n)^3}{V}Q_{tot}\Gamma \tag{6.5.55}$$

where τ_{nr} is the non-radiative scattering time, τ_{sp}^0 the free space spontaneous emission time, and $Q_{tot}^{-1} = Q_{ohm}^{-1}+Q_{rad}^{-1}+Q_{isb}^{-1}$. The approximation holds since $\tau_{nr} \ll \tau_{sp}$. The out-coupling efficiency of radiation is

$$\eta_{out} = \frac{1}{1+Q_{rad}/Q_{ohm}} \tag{6.5.56}$$

where the radiative quality factor is Q_{rad}, the ohmic quality factor Q_{ohm}. The total efficiency is then given by

$$\eta_{tot} = \frac{6}{\pi^2} \frac{Q_{tot}\Gamma}{(1 + Q_{rad}/Q_{ohm})} \frac{\tau_{nr}}{\tau_{sp}^0} \frac{(\lambda/2n)^3}{V} \tag{6.5.57}$$

The following examples illustrate such an emitter. The three LC resonators shown in Fig. 6.30 are compared; the planar resonator of the LC laser, a thin version of this resonator, and a 3D resonator. For an emitter active region at 1.5 THz, a spontaneous emission lifetime of $\tau_{sp}^0 = 30$ μs and a non-radiative lifetime of $\tau_{nr} = 5$ ps and a linewidth of $Q_{isb} = 4 \approx Q_{tot}$ is assumed. The table 6.1 summarizes the resonator properties and the computed efficiencies for terahertz generation. The planar and thin planar LC resonator lead to similar total efficiencies. A smaller volume results in a larger Purcell factor but also in a lower radiation efficiency. Actually the planar LC resonators have an efficiency that is barely above the internal efficiency of terahertz generation in the material without a cavity: $\tau_{nr}/\tau_{sp}^0 = 1.8$e-7. The small volume combined with a good radiation out-coupling of the 3D resonator results in an important increase of the total efficiency (a factor of ~ 100) in the present case. For the non-radiative lifetime a typical value at cryogenic temperatures has been assumed. At room temperature it would drop an order of magnitude, leading to 10 times smaller efficiencies. The expected emission is narrowband, with a linewidth in the order of 20-60 Ghz at a frequency of 1.5 THz. The predicted efficiency at cryogenic temperatures with the 3D LC resonator is about 1000 times larger than what has been measured on parabolic quantum wells [205] in agreement with a predicted Purcell factor of ~ 1000. A Purcell emitter at higher frequency,

LC resonator	planar	thin planar	3D
Q_{ohm}	51	27	97
Q_{rad}	189	1300	183
$V/(\lambda/2n)^3$	0.12	0.02	0.002
Γ	0.85	0.93	0.85
$\bar{F}_p$	17	113	1033
η_{int}	3.0e-6	2.0e-5	1.8e-4
η_{out}	2.1e-1	2.0e-2	3.5e-1
η_{tot}	6.5e-7	4.1e-7	6.4e-5

Table 6.1: Summary of the relevant resonator parameters and the computed efficiencies.

such as 3 or 4 THz, would take advantage of a larger Q_{isb} and a shorter τ_{sp}^0 and therefore larger efficiencies are expected.

Chapter 7

Conclusions and perspectives

In the first part, a bound-to-continuum bandstructure with split injector has been developed for low frequency terahertz QCLs. The main advantage of this bandstructure is the low intersubband absorption at the lasing frequency due to the minigap. The large scattering phase space due to the miniband results in a favorable condition for population inversion. A good injection efficiency at alignment bias is obtained due to the split-injector.

Based on the bound-to-continuum bandstructure a series of terahertz QCLs has been demonstrated, covering the frequency range from 2.1 to 1.2 THz. The latter is the lowest reported operation frequency of a terahertz QCL without a strong applied magnetic field. The relative injection efficiency of low frequency terahertz QCLs is studied via magneto-transport measurements. The limiting mechanism at low frequency is related to the low photon energy in combination with ionized impurity scattering. The smaller energy spacing between the upper and lower state leads to a decrease of the injection efficiency at alignment bias for lower frequencies. In the resonant tunneling picture the detuning from the parasitic injection channel is insufficient at injection resonance into the upper state. The coupling and broadening of the injector determine the width (or bias range) of the injection resonance. The low frequency designs with the modified doping profile suffer from ionized impurity scattering of the injector state. This leads to a strong non-selective leakage channel prior to the alignment bias. It results in a reduced dynamic range for the 1.5 and 1.3 THz laser, and in an unstable working point at the alignment bias of the designs aiming for 1.1 and 1.0 THz. However no indications for a drastic increase of the optical losses at lower frequency, nor a dramatic decrease of the internal quantum efficiency, that reflects the lifetimes, are found in the studied lasers. Future work aiming towards the 1 THz laser should address the doping issue. A doping profile similar to the lasers operating above 1.6 THz should be re-considered. To increase the injection selectivity, thicker injector barriers could be considered as well.

In the second part a deep sub-wavelength circuit based microcavity laser at 1.5 THz is demonstrated. The effective mode volume is among the smallest for electrically pumped lasers. The resonator consists of a planar capacitor-inductor resonant circuit. The resonator losses are dominated by ohmic losses. Simulations predict lower losses and at the same time significantly smaller effective mode volumes in LC resonators with a 3D inductor wire. Quantum electrodynamic effects are observed in the circuit based resonator. From the study of the emission characteristics of the circuit based laser, a Purcell factor in the order of $15 - 20$ is deduced. A series of modulation doped quantum wells is used as the active medium in a circuit based resonator for the observation of intersubband polaritons; the light-matter strong coupling regime. The circuit based laser is only one example of a whole class of applications opened by the circuit based resonator. The very high confinement of the electric field it enables makes it extremely attractive for emitters, detectors, modulators, and also for quantum optic studies.

Appendix A

Processing

A.1 Acid based etch solutions

Table A.1 summarizes the etch solutions used for sample fabrication and their approximate etching rates. All etch solutions are used at room temperature (21°C). The etch rate can depend on the age of the etchant, temperature, agitation in the etchant, shape of the structures, and size of the sample [85].

Etchant (acid)	Ratio	GaAs	$Al_{0.5}Ga_{0.5}As$	Nitride
Citric	$C_6H_8O_7$:H_2O:H_2O_2 (3:3:2)	30 μm/h	$< 0.3\mu$m/h	
Phosphoric	H_3PO_4:H_2O_2:H_2O (3:1:50)	80 nm/min	100 nm/min	
Sulphuric	H_2SO_4:H_2O_2:H_2O (1:8:13)	0.12 μm/s	0.12 μm/s	
Hydrofluoric	HF	0	500 nm/min	1 μm/min
Hydrofluoric	HF:H_2O (1:5)	0	0	0.3 μm/min

Table A.1: Etchants used for wet etching. Ratios are in volume, except for the citric acid based etchant, where the mass ratio is given. Etch rates are estimated and may depend on the etching conditions.

	spinning	baking	exposition	development
AZ1518	4000/4/40	1'/100°C	typ. 105 mJ, min. 90 mJ	MIF 726, 20"
MAN1420	5000/5/40	1.5'/95°C	972 mJ (split in 3 seq.)	MAD 533s, ∼ 2'

Table A.2: Recipes for the positive resist AZ1518 and the negative resist MAN1420.

	spinning	baking	expo	baking	flood expo	development
AZ5214	4000/4/40	1' 100°C	48 mJ	45" 120°C	≥200 mJ typ 250 mJ	MIF 726, ≥28" typ 35"

Table A.3: Recipes for the image reversal resist AZ5214. After the first baking the resist can be used as positive resist with the same parameters as the AZ1518.

A.2 Resists

Recipes for three resists are summarized in Table A.2 and A.3. AZ1518 is a positive, MAN1420 a negative and AZ5214 an image reversal resist. The latter can be used in the same way as the AZ1518, a fact that is very useful to remove resist on the sample edges in a first exposure–development step. The resist is turned into a negative resist after a shallow exposition, by an image reversal baking and a flood exposure. The exposition energy is specified per cm^2 and is converted into an exposure time by measuring the intensity of the mercury lamp. The recipes are based on the measured intensity at 405 nm. The Karl Süss MA6 mask aligner is used in the constant power mode for soft contact photolithography. During the development the sample is immersed parallel to the surface in the developer base, hold laterally by tweezers and moved back and forth with a typical frequency of 2 Hz.

A.3 Plasma processing

Reproducible plasma processing requires careful conditioning of the reaction chamber, especially in respect to contaminations. Therefore the processing steps described below are in general preceded by a *pump* - N_2 *purge* - *predeposition flow* conditioning and followed by a *pump* - N_2 *purge* cycle.

A.3.1 Nitride deposition

The Oxford Instruments PECVD 80+ system is used for nitride deposition based on plasma enhanced chemical vapor deposition (PECVD). For silicon nitride SiN_x deposition the sample is heated to 120°C. The process gas is 2.5%-$SiH_4(N_2)$: NH_3 : N_2 = (186:3:69), and the chamber pressure is regulated to 600 mTorr. Only HF power at 10 Watt is used. 70 min of deposition corresponds approximately to a 1.3 μm thick nitride layer.

A.3.2 Nitride etching

The Oxford Instruments RIE 80+ system is used for dry etching of the nitride by reactive ion etching (RIE). The etching process step is carried out at room temperature, the etching gas is CHF_3 : O_2 = (50:5) at 25 mTorr pressure. The RF generator delivers 55 Watt. 50 minutes of etching are sufficient to etch completely the nitride layer.

A.3.3 GaAs etching

The Oxford Instruments ICP 180 system is used for dry etching of the GaAs/ $Al_{0.1}Ga_{0.9}As$ active region by inductively coupled plasma etching (ICP). This step is the most critical one in terms of contamination. The ICP 180 system is equipped with a load lock system. Prior to the sample loading, the silicon carrier wafer is cleaned in an etch step at 20 mTorr and room temperature with typically 150 Watt of RF power and 1500 Watt of ICP power. The process gas is $SF_6 : O_2 = (40:40)$. After typically 10 minutes of etching the sample is loaded. Processing starts with a plasma oxygen etch step during 3 minutes to remove residuals of the resist mask. The etch parameters are 20°C, the valve position is on 10 degrees, 130 Watt of RF power, and 100 sccm of O_2. The GaAs dry etching is carried out at 60°C, 5 mTorr of pressure, 140 Watt RF power and 400 Watt ICP power. The process gas is Cl_2:Ar = (13:5). After typically $\leq$ 4 min 45 sec, the 8 μm thick active region is completely etched. Another plasma oxygen step, identical to the previous finishes the process.

Appendix B

Microstrip-lines

B.1 Propagation constant

The geometry of a microstrip-line is shown in figure B.1. A conducting strip of width W is on the top of a dielectric substrate that has a relative dielectric constant ε_r and a thickness h, and the bottom of the substrate is a conducting ground plane. An excellent reference for microstrip lines is given in [117]. The microstrip line can be characterized in the quasi-TEM approximation by an effective dielectric permittivity. The transmission characteristics are described by the effective dielectric constant ε_{re} and characteristic impedance Z_c. For thin conductors, the

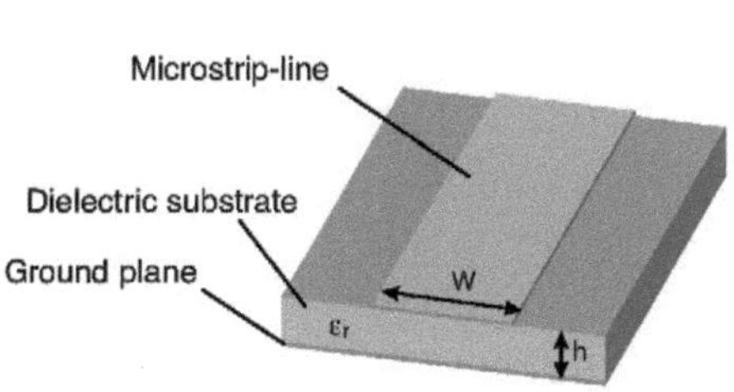

Figure B.1: Geometry of a microstrip-line. A conducting strip is separated by a dielectric substrate from a conducting ground plane.

closed-form expressions that provides an accuracy better than one percent are given as follows, where $\eta = 120\pi$ ohms is the wave impedance in free space.

For $W/h \leq 1$:

$$\begin{aligned} \varepsilon_{re} &= \frac{\varepsilon_r + 1}{2} + \frac{\varepsilon_r - 1}{2}\left\{\left(1 + 12\frac{h}{W}\right)^{-0.5} + 0.04\left(1 - \frac{W}{h}\right)^2\right\} \\ Z_c &= \frac{\eta}{2\pi\sqrt{\varepsilon_{re}}}\ln\left(\frac{8h}{W} + 0.25\frac{W}{h}\right) \end{aligned}$$

For $W/h \geq 1$:

$$\begin{aligned} \varepsilon_{re} &= \frac{\varepsilon_r + 1}{2} + \frac{\varepsilon_r - 1}{2}\left(1 + 12\frac{h}{W}\right)^{-0.5} \\ Z_c &= \frac{\eta}{\sqrt{\varepsilon_{re}}}\left\{\frac{W}{h} + 1.393 + 0.677\ln\left(\frac{W}{h} + 1.444\right)\right\}^{-1} \end{aligned}$$

The propagation constant in a microstrip line is $\gamma = \alpha + i\beta$ where α are the losses and β is given by $\beta = 2\pi/\lambda_g = 2\pi\sqrt{\varepsilon_{re}}/\lambda_0$ and λ_0 being the free space wavelength. A very simple expression allows to estimate the losses in microstrip lines:

$$\alpha = \frac{R_s}{Z_c W} \quad \text{(1/unit length)}$$

where Z_c is the characteristic impedance of the microstrip of the width W, and R_s the surface resistance, determined by the skin depth. It is given by

$$R_s = \sqrt{\frac{\omega\mu_0}{2\sigma}}$$

where σ is the conductivity, μ_0 the permeability of free space, and ω the angular frequency. Note that the here defined losses are based on the attenuation of the amplitude of the field. The waveguide losses in QCLs are defined on the attenuation of the power, therefore the corresponding waveguide losses are 2α.

B.2 Open Ends

At the open end of a microstrip-line, the fields do not stop abruptly but extend slightly further due to the effect of the fringing field. This effect can be modeled with an additional equivalent length of the transmission line Δl. Following formulas allow to compute Δl with an accuracy that is better than 0.2% for the range of $0.01 \leq w/h \leq 100$ and $\varepsilon_r \leq 128$ [117]

$$\frac{\Delta l}{h} = \frac{\xi_1 \xi_3 \xi_5}{\xi_4} \tag{B.2.1}$$

where h is the thickness of the substrat and ξ_i is given by

$$\xi_1 = 0.434907 \frac{\varepsilon_{re}^{0.81} + 0.26(W/h)^{0.8544} + 0.236}{\varepsilon_{re}^{0.81} - 0.189(W/h)^{0.8544} + 0.87} \tag{B.2.2}$$

$$\xi_2 = 1 + \frac{(W/h)^{0.371}}{2.35\varepsilon_r + 1} \tag{B.2.3}$$

$$\xi_3 = 1 + \frac{0.5274 \tan^{-1}[0.084(W/h)^{1.9413/\xi_2}]}{\varepsilon_{re}^{0.9236}} \tag{B.2.4}$$

$$\xi_4 = 1 + 0.037 \tan^{-1}[0.067(W/h)^{1.456}]\{6 - 5\exp[0.036(1 - \varepsilon_r)]\} \tag{B.2.5}$$

$$\xi_5 = 1 - 0.218 \exp(-7.5W/h) \tag{B.2.6}$$

Appendix C

Microcavity Rate equations

C.1 Solution of the rate equations

The QCL rate equations 6.4.30 - 6.4.32 are solved in the steady state regime to obtain a relation between the photon number p and the current density J. The calculations are performed for $\beta = 1$. The equations to solve are

$$\frac{dN_3}{dt} = \frac{J}{qL} - \frac{N_3}{\tau_3} - \frac{p(N_3 - N_2)}{\tau_{sp}} - \frac{N_3}{\tau_{sp}} \tag{C.1.1}$$

$$\frac{dN_2}{dt} = \frac{N_3}{\tau_{32}} - \frac{N_2}{\tau_2} + \frac{p(N_3 - N_2)}{\tau_{sp}} + \frac{N_3}{\tau_{sp}} \tag{C.1.2}$$

$$\frac{dp}{dt} = -\frac{p}{\tau_c} + \frac{pV(N_3 - N_2)}{\tau_{sp}} + \frac{VN_3}{\tau_{sp}} \tag{C.1.3}$$

N_3 and N_2 are eliminated and the current density as a function of the photon number writes

$$J = \frac{qL}{V\tau_c}\left[\frac{\tau_{sp}/\tau_3 + p\tau_2/\tau_3}{(p+1)/p - \tau_2/\tau_{32}} + p\right] \tag{C.1.4}$$

This equation can be solved for the photon number p. Following expression is obtained

$$p = \frac{\left(\frac{JV\tau_c\tau_{eff}}{qL} - \tau_3 - \tau_{sp}\right) + \sqrt{\left[\frac{JV\tau_c\tau_{eff}}{qL} - \tau_3 - \tau_{sp}\right]^2 + \frac{4JV\tau_c\tau_3(\tau_2+\tau_{eff})}{qL}}}{2(\tau_2 + \tau_{eff})} \tag{C.1.5}$$

where $\tau_{eff} = \tau_3(1 - \tau_2/\tau_{32})$ has been introduced to simplify the notation. The threshold current density J_{th} is obtained by requiring the net stimulated gain equal to the losses in Eq. C.1.3. The terms in $1/\tau_{sp}$ are neglected in Eq. C.1.1 and Eq. C.1.2.

$$J_{th} = \frac{qL}{V\tau_c} \frac{\tau_{sp} + \tau_3}{\tau_{eff} - \tau_3\tau_2/\tau_{sp}} \approx \frac{qL}{V\tau_c} \frac{\tau_{sp}}{\tau_{eff}} \tag{C.1.6}$$

The approximation makes use of the fact that in the LC laser $\tau_{sp}/\tau_3 \sim 10^6$. The same approximation is used to simplify Eq. C.1.5 and expressed with the threshold current density it writes as

$$p \approx \left[\left(\frac{J}{J_{th}} - 1\right) + \sqrt{\left(\frac{J}{J_{th}} - 1\right)^2 + 4\frac{J}{J_{th}} \frac{\tau_3}{\tau_{eff}} \frac{\tau_2 + \tau_{eff}}{\tau_{sp}}}\right] \frac{\tau_{sp}}{2(\tau_2 + \tau_{eff})} \tag{C.1.7}$$

Introducing the inversion ratio r, the saturation photon number p_{sat} and f Eq. C.1.7 is expressed as

$$p = \left((r-1) + \sqrt{(r-1)^2 + \frac{4rf}{p_{sat}}}\right) \frac{p_{sat}}{2} \tag{C.1.8}$$

where

$$r = J/J_{th} \tag{C.1.9}$$

$$p_{sat} = \tau_{sp}/(\tau_2 + \tau_{eff}) \tag{C.1.10}$$

$$f = \tau_3/\tau_{eff} \tag{C.1.11}$$

C.2 Approximation

Eq. C.1.8 can be approximated by the following expression, if $|r-1| > \delta$, therefore for all r except if r is close to unity.

$$p \approx [(r-1) + |r-1|]\frac{p_{sat}}{2} + \frac{rf}{|r-1|} \tag{C.2.12}$$

The condition on δ is $4rf/p_{sat} \ll \delta^2$. Since r and f are in the order of 1, and p_{sat} is in the order of 10^6 in the LC laser, $0.01 \leq \delta$. Two cases can be distinguished; below threshold ($r < 1$) and above threshold ($r > 1$). The approximate solutions are then

$$\text{if } r < 0.99: \qquad p \approx \frac{r}{1-r}f \tag{C.2.13}$$

$$\text{if } r > 1.01: \qquad p \approx (r-1)p_{sat} \tag{C.2.14}$$

C.3 Purcell factor fit in the threshold region

With the fit of the $p(r)$ curve the ratio p_{sat}/f is deduced, and the threshold current density is measured. This corresponds to two equations, with four unknown variables τ_3, τ_2, τ_{32}, τ_{sp}, where the latter contains the Purcell factor. The Purcell factor is expressed as a function of two parameters, e.g. τ_2 and τ_3. Two experimentally determined constants χ and φ are introduced

$$\frac{p_{sat}}{f} = \frac{\tau_{sp}}{\tau_3}\frac{\tau_{eff}}{\tau_2 + \tau_{eff}} = \varphi \tag{C.3.15}$$

$$\frac{J_{th} V \tau_c}{qL} = \frac{\tau_{sp}}{\tau_{eff}} = \chi \tag{C.3.16}$$

Those two equations are used to eliminate τ_{32}, then the Purcell factor is expressed using $F_p = \tau_{sp}^0/\tau_{sp}$.

$$F_p = \frac{2\tau_{sp}^0}{\varphi\tau_3}\left[1 + \sqrt{1 + 4\frac{\chi}{\varphi}\frac{\tau_2}{\tau_3}}\right]^{-1} \tag{C.3.17}$$

τ_{32} is also expressed as a function of the same variables

$$\tau_{32} = \frac{\tau_2}{1 - \frac{\varphi}{2\chi}\left[1 + \sqrt{1 + 4\frac{\chi}{\varphi}\frac{\tau_2}{\tau_3}}\right]} \tag{C.3.18}$$

The condition $\tau_{32} > \tau_3$ must be satisfied and restricts the possible set of τ_2 and τ_3 values.

Bibliography

[1] Siegel, P. Terahertz technology. *IEEE Trans. Microwave Theory Tech.* **50**, 910–928 (2002).

[2] Phillips, T. & Keene, J. Submillimeter astronomy. *Proc. IEEE* **80**, 1662–1678 (1992).

[3] Tonouchi, M. Cutting-edge terahertz technology. *Nat. Photonics* **1**, 97–105 (2007).

[4] Woolard, D., Brown, E., Pepper, M. & Kemp, M. Terahertz frequency sensing and imaging: A time of reckoning future applications? *Proc. IEEE* **93**, 1722–1743 (2005).

[5] Mittelman, D., Jacobsen, R. & Nuss, M. T-ray imaging. *IEEE J. Select. Topics Quantum Electron.* **2**, 679–692 (1996).

[6] Ferguson, B. & Zhang, X. Materials for terahertz science and technology. *Nature Mat.* **1**, 26–33 (2002).

[7] Asada, M., Suzuki, S. & Kishimoto, N. Resonant tunneling diodes for subterahertz and terahertz oscillators. *Jpn. J. Appl. Phys.* **47**, 4375–4384 (2008).

[8] Brown, E. *et al.* Oscillations up to 712 ghz in inas/alsb resonant-tunneling diodes. *Appl. Phys. Lett.* **58**, 2291–2293 (1991).

[9] Rodwell, M., Le, M. & Brar, B. Inp bipolar ics: Scaling roadmaps, frequency limits, manufacturable technologies. *Proc. IEEE* **96**, 271 (2008).

[10] Virginia diodes. http://www.virginiadiodes.com/.

[11] Gorshunov, B. *et al.* Terahertz bwo-spectroscopy. *Intern. J. Infrared Millim. Waves* **26**, 1217 (2005).

[12] Kozlov, G. & Volkov, A. *Millimeter and Submillimeter Wave Spectroscopy of Solids*, chap. Coherent source submillimeter wave spectroscopy, 51–110 (Springer-Verlag, 1998).

[13] Nagatsuma, T., Ito, H. & Ishibashi, T. High-power rf photodiodes and their applications. *Laser and Photonics Rev.* **3**, 123–137 (2009).

[14] Kawase, K., Shikata, J. & Ito, H. Terahertz wave parametric source. *J. Phys. D: Appl. Phys.* **35**, R1 (2002).

[15] Gaidis, M. *et al.* A 2.5-thz receiver front end for spaceborne applications. *IEEE Trans. Microwave Theory Tech.* **48**, 733–739 (2000).

[16] Brndermann, E. *Long Wavelength Infrared Semiconductor Lasers*, chap. Widely tunable far-infrared hot-hole semiconductor lasers, 279–343 (Wiley, Hoboken, NJ, 2004).

[17] Muravjov, A. *et al.* Injection-seeded internal-reflection-mode p-ge laser exceeds 10 w peak terahertz power. *J. Appl. Phys.* **103**, 083112 (2008).

[18] Williams, B., Kumar, S., Hu, Q. & Reno, J. High-power terahertz quantum-cascade lasers. *IEE Elect. Lett.* **42**, 89–90 (2006).

[19] Qin, Q., Williams, B., Kumar, S., Reno, J. & Hu, Q. Tuning a terahertz wire laser. *Nat. Photonics* **3**, 732–737 (2009).

[20] Kumar, S., Hu, Q. & Reno, L. 186 k operation of terahertz quantum-cascade lasers based on a diagonal design. *Appl. Phys. Lett.* **94**, 131105 (2009).

[21] Maulini, R. *Broadly tunable mid-infrared quantum cascade lasers for spectroscopic applications.* Ph.D. thesis, Institut de Physique, Université de Neuchâtel (2006).

[22] Faist, J., Capasso, F., Sirtori, C., Sivco, D. & Cho, A. Quantum cascade lasers. In Liu, H. & Capasso, F. (eds.) *Intersubband transitions in quantum wells: Physics and device applications II*, vol. 66, chap. 1, 1–83 (Academic Press, 2000).

[23] Lax, B. *Quantum Electronics, A Symposium* (Columbia University, New York, 1960).

[24] Cho, A. *Molecular Beam Epitaxy* (AIP Press, Woodbury, NW, 1994).

[25] Kazarinov, R. & Suris, R. Possibility of the amplification of electromagnetic waves in a semiconductor with a superlattice. *Sov. Phys. Semicond.* **5**, 707–709 (1971).

[26] Chang, L., Esaki, L. & Tsu, R. Resonant tunneling in semiconductor double barriers. *Appl. Phys. Lett.* **24**, 593–595 (1974).

[27] West, L. & Eglash, S. First observation of an extremely large-dipole infrared transition within the conduction band of a gaas quantum well. *Appl. Phys. Lett.* **46**, 1156–1158 (1985).

[28] Helm, M., England, P., Colas, E., DeRosa, F. & Allen, S. Intersubband emission from semiconductor superlattices excited by sequential resonant tunneling. *Phys. Rev. Lett.* **63**, 74–77 (1989).

[29] Faist, J. *et al.* Quantum cascade laser. *Science* **264**, 553–556 (1994).

[30] Faist, J. *et al.* Narrowing of the intersubband electroluminescent spectrum in coupled-quantum well heterostructures. *Appl. Phys. Lett.* **65**, 94–96 (1994).

[31] Beck, M. *et al.* Continuous wave operation of a mid-infrared semiconductor laser at room temperature. *Science* **295**, 301–305 (2002).

[32] Köhler, R. *et al.* Terahertz semiconductor-heterostructure laser. *Nature* **417**, 156–159 (2002).

[33] Alpes lasers. http://www.alpeslasers.com/.

[34] Pranalytica. http://www.pranalytica.com/.

[35] Daylight solutions. http://www.daylightsolutions.com/.

[36] Bai., Y., Slivken, S., Darvish, S. & Razeghi, M. Room temperature continuous wave operation of quantum cascade lasers with 12.5% wall plug efficiency. *Applied Physics Letters* **93**, 021103 (2008).

[37] Lyakh, A. *et al.* 3 w continuous-wave room temperature single-facet emission from quantum cascade lasers based on nonresonant extraction design approach. *Appl. Phys. Lett.* **95**, 141113 (2009).

[38] Cathabard, O., Teissier, R., Devenson, J., Moreno, J. & Baranov, A. Quantum cascade lasers emitting near 2.6 μm. *Appl. Phys. Lett.* **96**, 141110 (2010).

[39] Walther, C. *et al.* Quantum cascade lasers operating from 1.2 to 1.6 thz. *Appl. Phys. Lett.* **91**, 131122 (2007).

[40] Bastard, G. *Wave mechanics applied to semiconductor heterostructures* (Les éditions de physique, Les Ulis, France, 1988).

[41] Sirtori, C., Capasso, F., Faist, J. & Scandolo, S. Nonparabolicity and a sum rule associated with bound-to-bound and bound-to-continuum intersubband transitions in quantum wells. *Phys. Rev. B* **50**, 8663–8674 (1994).

[42] Nelson, D., Miller, R. & Kleinman, D. Band nonparabolicity effects in semiconductor quantum wells. *Phys. Rev. B* **35**, 7770–7773 (1987).

[43] Yariv, A. *Quantum Electronics* (John Wiley & Sons, Inc., 1989), 3rd edn.

[44] Rosencher, E. & Vinter, B. *Optoelectronics* (Cambridge University Press, 2002).

[45] Sirtori, C. *et al.* Resonant tunneling in quantum cascade lasers. *IEEE J. Quantum Electron.* **34**, 1722–1729 (1998).

[46] Kumar, S. & Hu, Q. Coherence of resonant-tunneling transport in terahertz quantum-cascade lasers. *Phys. Rev. B* **80**, 245316 (2009).

[47] Dupont, E., Fathololoumi, S. & Liu, H. Simplified density-matrix model applied to three-well terahertz quantum cascade lasers. *Phys. Rev. B* **81**, 205311 (2010).

[48] Kazarinov, R. & Suris, R. Electric and electromagnetic properties of semiconductors with a superlattice. *Sov. Phys. Semicond.* **6**, 120–131 (1972).

[49] Willenberg, H., Döhler, G. & Faist, J. Intersubband gain in a bloch oscillator and quantum cascade laser. *Phys. Rev. B* **67**, 085315 (2003).

[50] Terazzi, R., Gresch, T., Wittmann, A. & Faist, J. Sequential resonant tunneling in quantum cascade lasers. *Phys. Rev. B* **78**, 155328 (2008).

[51] Callebaut, H., Kumar, S., Williams, B., Hu, Q. & Reno, J. Analysis of transport properties of terahertz quantum cascade lasers. *Appl. Phys. Lett.* **83**, 207–209 (2003).

[52] Jirauschek, C., Scarpa, G., Lugli, P., Vitiello, M. & Scamarcio, G. Comparative analysis of resonant phonon thz quantum cascade lasers. *J. Appl. Phys.* **101**, 086109 (2007).

[53] Lee, S. & Wacker, A. Nonequilibrium green's function theory for transport and gain properties of quantum cascade structures. *Phys. Rev. B* **66**, 245314 (2002).

[54] Terazzi, R. & Faist, J. A density matrix model of transport and radiation in quantum cascade lasers. *New J Phys* **12**, 033045 (2010).

[55] Savic, I. *et al.* Density matrix theory of transport and gain in quantum cascade lasers in a magnetic field. *Phys. Rev. B* **76**, 165310 (2007).

[56] Scalari, G., Terazzi, R., Giovannini, M., Hoyler, N. & Faist, J. Population inversion by resonant tunneling in quantum wells. *Appl. Phys. Lett.* **91**, 032103 (2007).

[57] Vitiello, M. *et al.* Measurement of subband electronic temperatures and population inversion in thz quantum-cascade lasers. *Appl. Phys. Lett.* **86**, 111115 (2005).

[58] Vitiello, M. *et al.* Probing quantum efficiency by laser-induced hot-electron cooling. *Appl. Phys. Lett.* **94**, 021115 (2009).

[59] Lee, S. & Galbraith, I. Multisubband nonequilibrium electron-electron scattering in semiconductor quantum wells. *Phys. Rev. B* **55**, R16025 (1997).

[60] Ferreira, R. & Bastard, G. Evaluation of some scattering times for electrons in unbiased and biased single-and multiple-quantum-well structures. *Phys. Rev. B* **40**, 1074–1086 (1989).

[61] Cohen-Tannoudji, C., Diu, B. & Laloe, F. *Quantum Mechanics* (John Wiley & Sons, Inc., 1977).

[62] Williams, B. Terahertz quantum-cascade lasers. *Nat. Photonics* **1**, 517–525 (2007).

[63] Callebaut, H., Kumar, S., Williams, B., Hu, Q. & Reno, J. Importance of electron-impurity scattering for electron transport in terahertz quantum-cascade lasers. *Appl. Phys. Lett.* **84**, 645–647 (2004).

[64] Nelander, R. & Wacker, A. Temperature dependence of the gain profile for terahertz quantum cascade laseres. *Appl. Phys. Lett.* **92**, 081102 (2008).

[65] Smet, J., Fonstad, C. & Hu, Q. Intrawell and interwell intersubband transitions in multiple quantum wells for far-infrared sources. *J. Appl. Phys.* **79**, 9305–9320 (1996).

[66] Ando, T., Fowler, A. & Stern, F. Electronic properties of two-dimensional systems. *Rev. Mod. Phys.* **54**, 437–672 (1982).

[67] Unuma, T., Yoshita, M., Noda, T., Sakaki, H. & Akiyama, H. Intersubband absorption linewidth in gaas quantum wells due to scattering by interface roughness, phonons, alloy disorder, and impurities. *J. Appl. Phys.* **93**, 1586–1597 (2003).

[68] Unuma, T. *et al.* Effects of interface roughness and phonon scattering on intersubband absorption linewidth in a gaas quantum well. *Appl. Phys. Lett.* **78**, 3448–3450 (2001).

[69] Tsujino, S. *et al.* Interface-roughness-induced broadening of intersubband electroluminescence in p-sige and n-gainas/alinas quantum cascade structures. *Appl. Phys. Lett.* **86**, 062113 (2005).

[70] Wittmann, A., Bonetti, Y., Faist, J., Gini, E. & Giovannini, M. Intersubband linewidths in quantum cascade laser designs. *Appl. Phys. Lett.* **93**, 141103 (2008).

[71] Leuliet, A. *et al.* Electron scattering spectroscopy by a high magnetic field in quantum cascade lasers. *Phys. Rev. B* **73**, 085311 (2006).

[72] Weisbuch, C., Dingle, R., Gossard, A. & Wiegmann, W. Optical characterization of interface disorder in gaas-ga$_{1-x}$al$_x$as multi-quantum well structures. *Solid State Communications* **38**, 709 (1981).

[73] Runge, E. Excitons in semiconductor nanostructures. *Solid State Physics* **57**, 149 (2002).

[74] Becker, C., Vasanelli, A., Sirtori, C. & Bastard, G. Electron-longitudinal optical phonon interaction between landau levels in semiconductor heterostructures. *Phys. Rev. B* **69**, 115328 (2004).

[75] Fischer, M. *et al.* Scattering processes in terahertz ingaas/inalas quantum cascade lasers. *Appl. Phys. Lett.* **97**, 221114 (2010).

[76] Scalari, G. *et al.* Strong confinement in terahertz intersubband lasers by intense magnetic fields. *Phys. Rev. B* **76**, 115305 (2007).

[77] Vasanelli, A. *et al.* Role of elastic scattering machanisms in gainas/alinas quantum cascade lasers. *Appl. Phys. Lett.* **89**, 172120 (2006).

[78] Ando, T. Line width of inter-subband absorption in inversion layers: Scattering from charged ions. *J. Phys. Soc. Jpn* **54**, 2671 (1985).

[79] Lee, S. & Galbraith, I. Intersubband and intrasubband electronic scattering rates in semiconductor quantum wells. *Phys. Rev. B* **59**, 15796–15805 (1999).

[80] Lü, J. & Cao, J. Coulomb scattering in the monte carlo simulation of terahertz quantum-cascade lasers. *Appl. Phys. Lett.* **89**, 211115 (2006).

[81] Nelander, R. & Wacker, A. Temperature dependence and screening models in quantum cascade structures. *J. Appl. Phys.* **106**, 063115 (2009).

[82] Jirauschek, C., Matyas, A. & Lugli, P. Modeling bound-to-continuum terahertz quantum cascade lasers: The role of coulomb interactions. *J. Appl. Phys.* **107**, 013104 (2010).

[83] Taklo, M., Storas, P., Schjolberg-Henriksen, K., Hasting, H. & Jakobsen, H. Strong, high-yield and low-temperature thermocompression silicon wafer-level bonding with gold. *J. Micromech. Microeng.* **14**, 884–890 (2004).

[84] Williams, R. *Modern GaAs processing methods* (Artech House, Inc., 1990).

[85] Baca, A. G. & Ashby, C. I. H. (eds.) *Fabrication of GaAs Devices* (The Institute of Electrical Engineers, London, UK, 2005).

[86] Rochat, M. *Far-infrared Quantum Cascade Lasers*. Ph.D. thesis, Institut de Physique, Université de Neuchâtel (2002).

[87] Scalari, G. *Magneto-spectroscopy and development of terahertz quantum cascade lasers*. Ph.D. thesis, Institut de Physique, Université de Neuchâtel (2005).

[88] Tredicucci, A. *et al.* Terahertz quantum cascade lasers. *Physica E* **21**, 846–851 (2004).

[89] Scalari, G. *et al.* Far-infrared ($\lambda \simeq 87\mu$m) bound-to-continuum quantum-cascade lasers operating up to 90 k. *Appl. Phys. Lett.* **82**, 3165–3167 (2003).

[90] Ajili, L. *et al.* High power quantum-cascade lasers operating at $\lambda \sim 87$ and 130 μm. *Appl. Phys. Lett.* **85**, 3986–3988 (2004).

[91] Worrall, C. *et al.* Continuous wave operation of a superlattice quantum cascade laser emitting at 2 thz. *Optics Express* **14**, 171–181 (2006).

[92] Amanti, M. I. *et al.* Bound-to-continuum terahertz quantum cascade laser with a single-quantum-well phonon extraction/injection stage. *New Journal of Physics* **11**, 125022 (2009).

[93] Williams, B., Callebaut, H., Kumar, S., Hu, Q. & Reno, J. 3.4-thz quantum cascade laser based on longitudinal-optical-phonon scattering for depopulation. *Appl. Phys. Lett.* **82**, 1015–1017 (2003).

[94] Williams, B., Kumar, S., Qin, Q., Hu, Q. & Reno, J. Terahertz quantum cascade lasers with double-resonant-phonon depopulation. *Appl. Phys. Lett.* **88**, 261101 (2006).

[95] Luo, H. *et al.* Terahertz quantum-cascade lasers based on a three-well active module. *Appl. Phys. Lett.* **90**, 041112 (2007).

[96] Scalari, G. *et al.* Step well quantum cascade laser emitting at 3 thz. *Appl. Phys. Lett.* **94**, 041114 (2009).

[97] Kumar, S., Chan, C., Hu, Q. & Reno, J. Two-well terahertz quantum-cascade laser with direct intrawell-phonon depopulation. *Appl. Phys. Lett.* **95**, 141110 (2009).

[98] Scalari, G. *et al.* Broadband thz lasing from a photon-phonon quantum cascade structure. *Optics Express* **18**, 8043 (2010).

[99] Scalari, G., Hoyler, N., Giovannini, M. & Faist, J. Terahertz bound-to-continuum quanutm-cascade lasers based on optical-phonon scattering extraction. *Appl. Phys. Lett.* **86**, 181101 (2005).

[100] Kohler, R. *et al.* Terahertz quantum-cascade lasers based on an interlaced photon-phonon cascade. *Appl. Phys. Lett.* **84**, 1266–1268 (2004).

[101] Walther, C., Scalari, G., Faist, J., Beere, H. & Ritchie, D. Low frequency terahertz quantum cascade laser operating from 1.6 to 1.8 thz. *Appl. Phys. Lett.* **89**, 231121 (2006).

[102] Scalari, G. *et al.* Population inversion by resonant magnetic confinement in terahertz quantum-cascade lasers. *Appl. Phys. Lett.* **83**, 3453–3455 (2003).

[103] Scalari, G., Walther, C., Faist, J., Beere, H. & Ritchie, D. Electrically switchable, two-color quantum cascade laser emitting at 1.39 and 2.3 thz. *Appl. Phys. Lett.* **88**, 141102 (2006).

[104] Scalari, G. *et al.* Terahertz emission from quantum cascade lasers in the quantum hall regime: evidence for many body resonances and localization effects. *Phys. Rev. Lett.* **93**, 237403 (2004).

[105] Scalari, G. *et al.* Thz and sub-thz quantum cascade lasers. *Laser and Photonics Rev.* **3**, 45–66 (2009).

[106] Kumar, S., Williams, B., Hu, Q. & Reno, J. 1.9 thz quantum-cascade lasers with one-will injector. *Appl. Phys. Lett.* **88**, 121123 (2006).

[107] Kumar, S. Recent progress in terahertz quantum cascade lasers. *IEEE J. Select. Topics Quantum Electron.* **17**, 38 (2010).

[108] Pankove, J. *Optical processes in semiconductors* (Dover Publications, Inc., New York, US, 1971).

[109] Faist, J., Beck, M., Aellen, T. & Gini, E. Quantum cascade lasers based on a bound-to-continuum transition. *Appl. Phys. Lett.* **78**, 147–149 (2001).

[110] Kempa, K. *et al.* Intersubband transport in quantum wells in strong magnetic fields mediated by single- and two-electron scattering. *Phys. Rev. Lett.* **88**, 226803 (2002).

[111] Ulrich, J. *et al.* Terahertz-electroluminescence in a quantum cascade structure. *Physica B* **272**, 216–218 (1999).

[112] Kohen, S., Williams, B. & Hu, Q. Electromagnetic modeling of terahertz quantum cascade laser waveguides and resonators. *J. Appl. Phys.* **97**, 053106 (2005).

[113] Unterrainer, K. *et al.* Quantum cascade lasers with double metal-semiconductor waveguide resonators. *Appl. Phys. Lett.* **80**, 3060–3062 (2002).

[114] Adam, A. *et al.* Beam patterns of terahertz quantum cascade lasers with subwavelength cavity dimensions. *Appl. Phys. Lett.* **88**, 151105 (2006).

[115] Ordal, M., Bell, R., Alexander, R., Long, L. & Querry, M. Optical properties of fourteen metals in the infrared and far infrared: Al, co, cu, au, fe, pb, mo, ni, pd, pt, ag, ti, v, and w. *Appl. Opt.* **24**, 4493–4499 (1985).

[116] Ordal, M., Bell, R., Alexander, R., Long, L. & Querry, M. Optical properties of au, ni, and pb at submillimeter wavelengths. *Appl. Opt.* **26**, 744–752 (1987).

[117] Hong, J.-S. & Lancaster, M. J. *Microstrip Filters for RF/Microwave Applications* (John Wiley & Sons, Inc., 2001).

[118] Blaser, S. *et al.* Characterization and modeling of quantum cascade lasers based on photon-assisted tunneling transition. *IEEE J. Quantum Electron.* **37**, 448–455 (2001).

[119] Faist, J. *et al.* Laser action by tuning the oscillator strength. *Nature* **387**, 777–782 (1997).

[120] Sze, S. *Modern semiconductor device physics* (John Wiley & Sons, Inc., New-York, 1998).

[121] Benz, A. *et al.* Influence of doping on the performance of terahertz quantum-cascade lasers. *Appl. Phys. Lett.* **90**, 101107 (2007).

[122] Pfeiffer, L., Schubert, E., West, K. & Magee, C. Si dopant migration and the algaas/gaas inverted interface. *Appl. Phys. Lett.* **58**, 2258 (1991).

[123] Wiedner, M. *et al.* First observations with condor, a 1.5 thz heterodyne receiver. *Astron. Astrophys.* **454**, L33 (2006).

[124] de Graauw, T., Helmich, F., Philips, T. & et al., J. S. The herschel-heterodyne instrument for the far-infrared (hifi). *Astronomy and Astrophysics* **518**, L6 (2010).

[125] Hübers, H. *et al.* Heterdyne receiver for 3-5 thz with hot-electron bolometer mixer. *Millimter and Submillimeter Detectors for Astronomy II.* **5498**, 579 (2004).

[126] Philipp, M., Graf, U., Wagner-Gentner, A., Rabanus, D. & Lewen, F. Compact 1.9 thz bwo local-oscillator for the great heterodyne receiver. *Infrared Physics and Technology* **51**, 54 (2007).

[127] Stratospheric observatory for infrared astronomy. http://www.sofia.usra.edu.

[128] Graf, U. *et al.* Great: the german first light heterodyne instrument for sofia. *Proc. of SPIE* **6678**, 66780K (2007).

[129] Rabanus, D. *et al.* Phase locking of a 1.5 terahertz quantum cascade laser and use as a local oscillator in a heterodyne heb receiver. *Optics Express* **17**, 1159 (2009).

[130] Blank, A. & Feng, S. Suppression of intersubband nonradiative transitions by magnetic field in quantum well laser devices. *J. Appl. Phys.* **74**, 4795–4797 (1993).

[131] Canali, L., Lazzarino, M., Sorba, L. & Beltram, F. Stark-cyclotron resonance in a semiconductor superlattice. *Phys. Rev. Lett.* **76**, 3618–3621 (1996).

[132] Raikh, M. & Shabazyan, T. Magnetointersubband oscillations of conductivity in a two-dimensional electronic system. *Phys. Rev. B* **49**, 5531–5540 (1994).

[133] Ferreira, R. Resonances in the hopping probability between flexible quantum dots: the case of superlattices under parallel electric and magnetic field. *Phys. Rev. B* **43**, 9336–9338 (1991).

[134] Ulrich, J., Zobl, R., Unterrainer, K., Strasser, G. & Gornik, E. Magnetic-field-enhanced quantum-cascade emission. *Appl. Phys. Lett.* **76**, 19–21 (2000).

[135] Blaser, S., Rochat, M., Beck, M., Hofstetter, D. & Faist, J. Terahertz intersubband emission in strong magnetic fields. *Appl. Phys. Lett.* **81**, 67–69 (2002).

[136] Alton, J. *et al.* Magnetic field in-plane quantization and tuning of population inversion in a thz superlattice quantum cascade laser. *Phys. Rev. B* **68**, 081303 (2003).

[137] Tamosiunas, V. *et al.* Terahertz quantum cascade lasers in a magnetic field. *Appl. Phys. Lett.* **83**, 3873–3875 (2003).

[138] Wade, A. *et al.* Magnetic-field-assisted terahertz quantum cascade laser operating up to 225 k. *Nat. Photonics* **3**, 41–45 (2008).

[139] Laperne, N. *et al.* Inter-landau level scattering and lo-phonon emission in terahertz quantum cascade lasers. *Appl. Phys. Lett.* **91**, 062102 (2007).

[140] Amanti, M. *et al.* Bound-to-continuum terahertz quantum cascade laser with a single-quantum-well phonon extraction/injection stage. *New J Phys* **11**, 125022 (2009).

[141] Chassagneux, Y. *et al.* Terahertz microcavity lasers with subwavelength mode volumes and thresholds in milliampere range. *Appl. Phys. Lett.* **90**, 091113 (2007).

[142] Park, H. *et al.* Electrically driven single-cell photonic crystal laser. *Science* **305**, 1444–1447 (2004).

[143] Scheuer, J., Green, W., DeRose, G. & Yariv, A. Lasing from a circular bragg nanocavity with an ultrasmall modal volume. *Appl. Phys. Lett.* **86**, 251101 (2005).

[144] Nozaki, K. & Baba, T. Laser characteristics with ultimate-small modal volume in photonic crystal slab point-shift nanolasers. *Appl. Phys. Lett.* **88**, 211101 (2006).

[145] Painter, O. *et al.* Two-dimensional photonic band-gap defect mode laser. *Science* **284**, 1819–1821 (1999).

[146] Hill, M. *et al.* Lasing in metallic-coated nanocavities. *Nat. Photonics* **1**, 589–594 (2007).

[147] Noginov, M. *et al.* Demonstration of a spaser-based nanolaser. *Nature* **460**, 1110–1112 (2009).

[148] Walther, C., Scalari, G., Amanti, M., Beck, M. & Faist, J. Microcavity laser oscillating in a circuit-based resonator. *Science* **327**, 1495–1497 (2010).

[149] Coccioli, R., Boroditsky, M., Kim, K., Samii, Y. & Yablonovitch, E. Smallest possible electromagnetic mode volume in a dielectric cavity. *IEE Proc. Optoelectron.* **145**, 391–397 (1998).

[150] Fasching, G. *et al.* Subwavelength microdisk and microring terahertz quantum-cascade lasers. *IEEE J. Quantum Electron.* **43**, 687 (2007).

[151] Dunbar, L. *et al.* Small optical volume terahertz emitting microdisk quantum cascade lasers. *Appl. Phys. Lett.* **90**, 141114 (2007).

[152] Oulton, R. *et al.* Plasmon lasers at deep subwavelength scale. *Nature* **461**, 629–632 (2009).

[153] Kühn, S., Hakanson, U., Rogobete, L. & Sandoghdar, V. Enhancement of single-molecule fluorescence using a gold nanoparticle as an optical nanoantenna. *Phys. Rev. Lett.* **97**, 017402 (2006).

[154] Engheta, N. Circuits with light at nanoscales: Optical nanocircuits inspired by metamaterials. *Science* **317**, 1698–1702 (2007).

[155] Linden, S. *et al.* Magnetic response of metamaterials at 100 terahertz. *Science* **306**, 1351 (2004).

[156] Smith, D., Pendry, J. & Wiltshire, M. Metamaterials and negative refractive index. *Science* **305**, 788–792 (2004).

[157] Veselago, V. G. *Sov. Phys. USPEKHI* **10**, 509 (1968).

[158] Smith, D., Padilla, W., Vier, D., Nasser, S. & Schultz, S. Composite medium with simultaneously negative permeability and permittivity. *Phys. Rev. Lett.* **84**, 4184 (2000).

[159] Pendry, J., Holden, A., Robbins, D. & Stewart, W. Magnetism from conductors and enhanced nonlinear phenomena. *IEEE Trans. Microwave Theory Tech.* **47**, 2075–2084 (1999).

[160] Yen, T. *et al.* Terahertz magnetic response from artificial materials. *Science* **303**, 1494 (2004).

[161] Padilla, W., Taylor, A., Highstrete, C., Lee, M. & Averitt, R. Dynamical electric and magnetic metamaterial response at terahertz frequencies. *Phys. Rev. Lett.* **96**, 107401 (2006).

[162] Wegener, M. *et al.* Toy model for plasmonic metamaterial resonances coupled to two-level system gain. *Optics Express* **16**, 19785 (2008).

[163] Gansel, J. *et al.* Gold helix photonic metamaterial as broadband circular polarizer. *Science* **325**, 1513 (2009).

[164] Comsol is a multiphysics modeling and simulation software. http://www.comsol.com/.

[165] Jin, J. *The Finite Element Method in Electromagnetics* (John Wiley & Sons, Inc., 2002), 2nd edn.

[166] Yahjian, A. & Best, S. Impedance, bandwidth, and q of antennas. *IEEE Trans. Antennas Propag.* **53**, 1298 (2005).

[167] Yahjian, A. Improved formulas for the q of antennas with highly lossy dispersive materials. *IEEE Antennas Wireless Propag. Lett.* **5**, 365 (2006).

[168] Pozar, D. M. *Microwave Engineering* (John Wiley & Sons, Inc., 2005), third edition edn.

[169] Lide, D. R. *CRC Handbook of Chemistry and Physics*, chap. Electrical Resistivity of Pure Metals (2008), 88th edn.

[170] Sze, S. M. *Physics of Semiconductor Devices* (John Wiley & Sons, Inc., 1981), 2nd edn.

[171] Zhang, H., Dunbar, L., Scalari, G., Houdré, R. & Faist, J. Terahertz photonic crystal quantum cascade lasers. *Optics Express* **15**, 16818–16827 (2007).

[172] Haroche, S. & Kleppner, D. Cavity quantum electrodynamics. *Phys. Today* **42**, 24 (1989).

[173] Purcell, E. Spontaneous emission probabilities at radio frequencies. *Phys. Rev.* **69**, 681 (1946).

[174] Gerard, J. *et al.* Enhanced spontaneous emission by quantum boxes in a monolithic optical microcavity. *Phys. Rev. Lett.* **81**, 1110–1114 (1998).

[175] Yuan, Z. *et al.* Electrically driven single-photon source. *Science* **295**, 102 (2002).

[176] Neve, H. *et al.* Recycling of guided mode light emission in planar microcavity light emitting diodes. *Appl. Phys. Lett.* **70**, 799 (1997).

[177] Kavokin, A. V., Baumberg, J. J., Malpuech, G. & Laussy, F. P. *Microcavities* (Oxford University Press Inc., New York, 2007).

[178] Martini, F. & Jacobovitz, G. Anomalous spontaneous-stimulated-decay phase transition and zero-threshold laser action in a microscopic cavity. *Phys. Rev. Lett.* **60**, 1711 (1988).

[179] Yokoyama, H. *et al.* Controlling spontaneous emission and threshold-less laser oscillation with optical microcavities. *Optical and Quantum Electronics* **24**, S245–S272 (1992).

[180] Gérard, J. Solid-state cavity-quantum electrodynamics with self-assembled quantum dots. *Topics Appl. Phys.* **90**, 269 (2003).

[181] Haroche, S. *Fundamental systems in quantum optics* (Elsevier Science Publishers, 1992).

[182] Todorov, Y., Sagnes, I., Abram, I. & Minot, C. Purcell enhancement of spontaneous emission from quantum cascades inside mirror-grating metal cavities at thz frequencies. *Phys. Rev. Lett.* **99**, 223603 (2007).

[183] Björk, G. & Yamamoto, Y. Analysis of semiconductor microcavity lasers using rate equations. *IEEE J. Quantum Electron.* **27**, 2386–2396 (1991).

[184] Balanis, C. A. *Antenna Theory* (John Wiley & Sons, Inc., 2005).

[185] Wheeler, H. Fundamental limitations of small antennas **35**, 1479–1484 (1947).

[186] Chu, L. Physical limitations of omni-directional antennas. *J. Appl. Phys.* **10**, 1163 (1948).

[187] McLean, J. S. A re-examination of the fundamental limits on the radiation q of electrically small antennas. *IEEE Trans. Antennas Propag.* **44**, 672 (1996).

[188] Johnson, P. & Christy, R. Optical constants of the noble metals. *Phys. Rev. B* **6**, 4370 (1972).

[189] Gao, J. R. *et al.* Terahertz heterodyne receiver based on a quantum cascade laser and a superconducting bolometer. *Applied Physics Letters* **86**, 244104 (2005).

[190] Semenov, A. *et al.* Superconducting hot-electron bolometer mixer for terahertz heterodyne receivers. *Applied Superconductivity, IEEE Transactions on* **13**, 168 – 171 (2003).

[191] Klopfenstein, R. W. A transmission line taper of improved design. *Proceedings of the IRE* **44**, 31 (1956).

[192] Chen, H. *et al.* Active terahertz metamaterial devices. *Nature* **444**, 597 (2006).

[193] Weisbuch, C., Nishioka, M., Ishikawa, A. & Arakawa, Y. Observation of the coupled exciton-photon mode splitting in a semiconductor quantum microcavity. *Phys. Rev. Lett.* **69**, 3314–3317 (1992).

[194] Khitrova, G., Gibbs, H., Jahnke, F., Kira, M. & Koch, S. Nonlinear optics of normal-mode-coupling semiconductor microcavities. *Rev. Mod. Phys.* **71**, 1591 (1999).

[195] Rapaport, R. *et al.* Negatively charged quantum well polaritons in a gaas/alas microcavity: An analog of atoms in a cavity. *Phys. Rev. Lett.* **84**, 1607 (2000).

[196] Saba, M. *et al.* Crossover from exciton to biexciton polaritons in semiconductor microcavities. *Phys. Rev. Lett.* **85**, 385 (2000).

[197] Tredicucci, A. *et al.* Controlled exciton-photon interaction in semiconductor bulk microcavities. *Phys. Rev. Lett.* **75**, 3906 (1995).

[198] Kelkar, P. *et al.* Excitons in a ii-vi semiconductor microcavity in the strong-coupling regime. *Phys. Rev. B* **52**, 5491 (1995).

[199] Dini, D., Koheler, R., Tredicucci, A., Biasiol, G. & Sorba, L. Microcavity polariton splitting of intersubband transitions. *Phys. Rev. Lett.* **90**, 116401 (2003).

[200] Sapienza, L. *et al.* Electrically injected cavity polaritons. *Phys. Rev. Lett.* **100**, 136806 (2008).

[201] Todorov, Y. *et al.* Strong light-matter coupling in subwavelength metal-dielectric microcavities at terahertz frequencies. *Phys. Rev. Lett.* **102**, 18640 (2009).

[202] Geiser, M. *et al.* Strong light-matter coupling at terahertz frequencies at room temperature in electronic lc resonators. *Appl. Phys. Lett.* **97**, 191107 (2010).

[203] Schneider, H. & Liu, H. *Quantum Well Infrared Photodetectors* (Springer, 2007).

[204] Graf, M. *et al.* Terahertz range quantum well infrared photodetector. *Appl. Phys. Lett.* **84**, 475–477 (2004).

[205] Ulrich, J. *et al.* Temperature dependence of far-infrared electroluminescence in parabolic quantum wells. *Appl. Phys. Lett.* **74**, 3158–3160 (1999).

MIX
Papier aus verantwortungsvollen Quellen
Paper from responsible sources
FSC® C105338

Printed by Books on Demand GmbH, Norderstedt / Germany